MW00934578

Names, Numbers, and Network Solutions

The Monetization of the Internet

J. Robert Beyster
with
Michael A. Daniels

Foreword by Vinton G. Cerf

The Foundation for Enterprise Development, La Jolla, CA

For more information about FED Productions, and our services and products, visit our web site at www.fed.org.

Library of Congress Cataloging-in-Publication Data:

ISBN-10: **1482077353**

ISBN-13: **9781482077353**

Library of Congress Control Number: **2013905458**

Printed in the United States of America.

Book Cover Design: Jamie Dickerson, The Foundation for Enterprise Development

CONTENTS

FOREWORD

I knew Bob Beyster at a time when SAIC was growing rapidly in size and scope. The book you are reading holds a particular significance for me, not only for the lessons that Bob and his coauthor, Mike Daniels, so clearly articulate, but also for the history they carefully render about a very important period in the history of the Internet. The story of the Internet's odyssey from military research to academic instrument to global infrastructure includes some critical points at which private sector engagement has been critical. Among these is access to the Internet to the general public and the subsequent transfer of US government–funded services to private sector operation. Whether the private sector operation was for-profit or not-for-profit, a business model was invariably needed to defray costs. If the private sector can be credited with anything, it is the remarkable ability to recognize and explore a wide range of business models to achieve sustainability. In the case of the Internet, this also has required an ability to deal with explosive exponential growth in demand in many dimensions. The story of the transition of the domain name system and Internet numerical address allocation from government-subsidized operation to private sector funding is captured well in this book, and is a helpful addition to the historical record.

If there is a hero to be highlighted in this part of the story, it has to be Mark Kosters who somehow managed to keep things going with inadequate infrastructure and what sounds like no sleep and no time off for eons. I had not appreciated, until reading this book, how close the Domain Name System came to collapse in consequence of the exponential growth in demand—not only for domain names but also their resolution into Internet addresses. Mark earned a place in my pantheon of less-recognized heroes, thanks to this documentation by Bob and Mike.

Among the most important realizations for me, as I read this gripping story, is the observation that the loose coupling of the institutions and networks that make up the Internet ecology is key to its resilience. There are myriad organizations that make up the Internet's ecosystem, and their voluntary collaboration and cooperation gives the system its vitality and ability to adapt to new conditions, technologies, and applications. Nowhere is this more evident as the authors trace the path of the important domain name and address allocation functions from the 1985 period but most especially from 1995-2000 during which the "dot-boom" took the scale and diversity of the Internet to new heights. Of course, this same time period produced some unbelievable and unsustainable valuations of Internet-related businesses. Investment mania led to massive amounts of capital, often going into companies whose business model seemed to be "we are somehow associated with the exploding Internet." Some of the founders did not seem to distinguish between capital and income and were surprised when the capital ran out. That's another story.

The SAIC story, on the other hand, has been a steady attention to opportunity, growth, and judicious risk-taking. One can discern the elements of the SAIC philosophy in the nuggets found throughout the book, which is worth reading for that reason alone, although I found it a very useful timeline of events during a very energetic period in the Internet's development and its transition from a research, academic, and military experiment to a growing global infrastructure that continues to astonish, with new applications arriving on what seems to be a daily basis.

Internet historians will welcome this addition to the spotty documentation of the Internet's story, and I am among them.

Vinton G. Cerf
McLean, Virginia

Preface

Some say timing is everything, and that was certainly the case when the company I founded—La Jolla, California-based Science Applications International Corporation (SAIC)—acquired Network Solutions, Inc. (NSI), in 1995. NSI was attractive to SAIC because of its well-established government telecommunications and computer network consulting business, and also because of its top-notch roster of commercial (nongovernment) clients, including AT&T and NationsBank.

But there was one more thing that NSI had, and it turned out to make all the difference in the world: a $4.2 million, five-year-nine-month cooperative agreement with the National Science Foundation (NSF) to provide InterNIC registration services. This meant, in short, that NSI controlled Internet domain name registration. If you wanted to establish a .com website at the time, then you had to get your domain name from NSI.

We bought NSI for $4.7 million just when the Internet was about to go through a growth spurt that was unprecedented in the history of business. As a result, our investment grew far beyond our greatest expectations, culminating in the sale of NSI to VeriSign in March 2000 for the final price of approximately $19.3 billion.[1]

While in hindsight this acquisition would seem to be a particularly brilliant business decision, in reality the pathway to success for Network Solutions was not known in advance, it was not linear, and it was anything but easy. When we took over NSI, upgrades to the company's Internet infrastructure had been underfunded for years, and basic security protections were lacking or completely nonexistent. In addition, NSI was not prepared for the fast-rising flood of domain name registrations that inundated the company within a year or two of the acquisition. Many members of the closely knit Internet community were watching NSI closely once the community figured out that this new gatekeeper intended to turn the registration of domain names into a business. And if that weren't enough, many different agencies and departments in the government at one time or another wanted to gain control over the Internet. Some wanted to regulate it, while others wanted to approve or disapprove it, take credit for success, or condemn our alleged shortcomings in administering the domain name system.

Suddenly, the Internet, which was on only a few Washington, DC radar screens prior to the 1990s, attracted the attention of many federal agencies including the FCC, the Commerce Department, and even the White House.

SAIC was in uncharted waters, but we knew that we had to move forward—our responsibility to enable and grow this increasingly important public asset required it. From a company standpoint, we were doing our best to manage a fast-growing business while meeting the interests of many Internet-involved individuals and institutions, as well as our employee-owners. On January 1, 1995, there were approximately 71,000 domain names in the entire world. By January 2005—just ten years later—the number of domain names had mushroomed to more than 46 million.[2] How we got there, how we got through it, and the lessons we learned along the way are what this book is all about.

You can invest much time and money in planning for the future, only to have the natural course of events overwhelm all your best-laid plans. And sometimes, when you make no plans at all, the most remarkable business opportunities will unexpectedly land on your doorstep.

In the many reports on the Internet's origins and rapid growth, the story of the domain name business—and specifically, Network Solutions' role in it—has not been addressed in much depth. Its evolutionary path was as much a surprise to the visionaries working in other areas of the Internet as it was to us. We've collected their views about Network Solutions to help provide the context of what the leading men and women in business, government, and academia were thinking about the company, monetizing the Internet, the scale of technical problems, and the relevance of this apparently obscure part of the Internet. Because of my roles at both SAIC and NSI, I have been able to gain an unprecedented degree of access to those people who made the Internet what it is today.

Being a part of a company that had such a major impact on the growth of the Internet is one of the most rewarding experiences of my entire life. I'm sure this is also the case for many others who took part in Network Solutions and for all the others who have created or grown an Internet business with substantial reach and impact. It is my hope that in these pages you too will gain a sense of what we at SAIC and NSI experienced during this time — the surprises and the challenges — and how this experience led to today's domain name system. It is also my hope that you will learn from our mistakes and from our successes.

Many times during the initial commercialization phase of exponential growth—approximately between the years 1995 and 2000—the Internet could have collapsed from its own weight, becoming a victim of its success. That it didn't is a tribute to the men and women whose vision and hard work brought us to where we are today. That it won't collapse from its own weight during the future exponential growth from mobile communications, always-on broadband Internet connections, voice over IP (VoIP), cloud computing, and much more to come will be a tribute to those men and women who will work hard to understand and solve critical problems before they become real-time crises.

What the Internet is today and what it will be tomorrow is the result of the efforts of many unsung operational heroes whose job it is to ferret out the strengths and frailties of technology, find common ground across the political spectrum, and serve the highest standards of customer and public service.

This story of Network Solutions offers lessons that remain relevant to entrepreneurs and policymakers even today. The Internet's infrastructure continues to quickly grow and transform, and new applications—including social networking sites such as Facebook, LinkedIn, and Twitter, along with iPads, iPhones, Google TV, Android, and new hardware and software applications that we have yet to invent—continue to offer unlimited opportunities to innovate and profit.

Although we learned a variety of lessons at SAIC from our brief, five-year ownership of Network Solutions, they resonated within the company for many years after. Here are four of the most important lessons we learned:

1. Government and Industry must work together to create massive innovation. The US government seeds and the private sector

operationalizes while creating jobs, innovation, and wealth. If we do not rededicate our nation to this, we will lose in the future.

2. Business is a good and positive force, and it needs to be supported and nurtured in our nation. A strong and vital and innovative private sector leads a nation to prosperity. Without it we face economic decline.

3. If you want to live in a fast-paced, challenging, exciting, and rewarding world, then the technology field is the place to be. This is always the future—it's where we create the future for everyone on earth.

4. Be honest, be ethical, work hard, and never give up. These are the keys to success, and they are what we always tried to do at SAIC and Network Solutions.

In the case of Network Solutions, some very talented and dedicated individuals did their part in forever changing the course of the domain name system and the Internet by helping to transition it from an obscure, government-run computer network into a new kind of global communications medium that rivals in importance and global impact such inventions as the telephone, radio, and television. I am personally quite proud of the role SAIC played in the Network Solutions story and in helping to improve and strengthen the infrastructure that kept the Internet running through a period of remarkable growth.

But we wouldn't have acquired Network Solutions—and we would not have played a key role in the commercialization of the Internet—if it were not for an SAIC employee who was keeping a very close eye on this emerging industry and carefully looking for opportunities. That employee was my coauthor Mike Daniels. You will read more about his role in identifying and tracking the industry and Network Solutions later in this book.

Every business needs someone like Mike—a person who is constantly scanning the horizon for new opportunities, and then who is willing to put his own neck on the line to push an organization's decision makers to see the merits in pursuing them.

Dr. J. Robert Beyster
La Jolla, California

PART ONE

SNAPSHOT OF AN INDUSTRY

I t's not often that an entirely new industry is created, but this was the case with the commercialization of the Internet in the 1990s. As ubiquitous as the Internet is in our lives today—with our increasing dependence on Internet-enabled e-mail, text and video messaging, streaming video, websites, blogs, and a constantly evolving array of other web apps—it's hard to imagine that the web has been around for little more than twenty years, and it didn't gain widespread usage until only about fifteen years ago.

The rapid growth of the Internet industry is unprecedented in recent times, and it has cycled through a number of booms and busts—and booms again. The story of Network Solutions is a microcosm of the larger story of the Internet industry, and it, too, endured its own cycle of booms and busts.

But what is it that determines when and where a new industry will emerge? And of perhaps even greater interest, how do business leaders decide which ones will be the winners, which ones will be the losers, and when they should jump in with their own offerings of new or improved products or services? While hindsight is golden, there are a few lessons to consider when it comes to the emergence of the Internet industry, and they are instructive for technology and other entrepreneurs, business owners, and leaders who hope to ride some future business wave to success.

Every great business success begins by creating or fulfilling the needs or desires of a large number of customers—there must be an unanswered, latent desire or a product or service necessity that someone or some company is willing to pay for. Even better, customer needs should be or should become urgent—at or near the top of the target list of priorities. Think of the urgency with which customers seek their morning coffee fix in a business originated by Starbucks. Now customers seek their coffee and Internet fixes simultaneously.

In the case of the Internet, it wasn't clear at first that there was a need at all on the part of most prospective customers, much less that this need was an urgent one. Before the Internet was commercialized, it was simply a worldwide computer network that allowed users (mostly government or academic computer scientists and researchers) to communicate with one another. However, as the Internet became increasingly commercialized during the 1990s, people and companies began to realize that accessing, leveraging—and even commanding—the Internet space was an increasingly urgent need.

This is exactly when the radar of savvy entrepreneurs and business leaders became alert to the emerging Internet industry, and when they began to develop new products and services to meet the urgent needs of their potential customers—preferably in a way that is better, more efficient, and less expensive than the competition. In the case of Network Solutions, that product was domain names. And because of the exclusive nature of its contract with the US government as the sole supplier of .com, .net, and .org domain names to the world, it was the only business that could meet the needs of the emerging Internet industry. When we acquired the company in 1995, we soon found that this new industry exceeded every growth estimate and target—with no limits in sight.

In 2011, the world population reached an estimated total of seven billion people. Of this total, it is estimated that more than two billion people around the world use the Internet. That's almost 30 percent of the earth's population. Considering that just thirty-five years ago the number of Internet users could have probably fit into a large lecture hall, it wouldn't surprise me to learn that no industry in the history of mankind has grown as fast as or faster than the Internet.

In this brief part, we'll take a look at the unprecedented growth of the Internet and introduce the five themes that comprise the backbone of this book:

- Be ready to exploit opportunities
- Build a solid financial basis
- Bring the best talent across diverse capabilities
- Engage with government and in the political process
- Know when to move on.

Small Business, Big Impact

The Internet will win because it is relentless. Like a cannibal, it even turns on its own. Though early portals like Prodigy and AOL once benefited from their first-mover status, competitors surpassed them as technology and consumer preferences changed. But like Shiva, the Internet creates as well as destroys. Social networks, search advertising, and cloud computing are multibillion dollar industries that didn't exist ten years ago. They are products of the same force that has rendered the postal service's core business obsolete. Today, they attract capital, spawn innovation, and employ millions; and tomorrow Vint Cerf, or someone in a garage, or the Internet gods will come up with an idea that takes it all away.

— John Sununu, Former New Hampshire senator[3]

This book is the story of a small business that helped change the world. Beyond that, this book is a narrative from entrepreneurs who lived through a major public-private sector commercialization experience, to entrepreneurs today and tomorrow to consult as they chart their own courses for changing the world.

The business—Network Solutions, Inc., headquartered in the northern Virginia suburbs of Washington, DC—helped to shepherd the growth of the Internet (specifically, the World Wide Web, or the web for short) at a critical time in its evolution and opened the way for billions of dollars of global commerce. And not only did NSI shepherd the web's growth, the company was instrumental in assuring that people, businesses, and other organizations all around the world were able to access and use the web twenty-four hours a day, seven days a week. If this small business failed (and as you will learn, sometimes it nearly did), the web itself failed.

But the company did for the most part succeed, thanks to the efforts of a dedicated group of very smart and talented men and women.

Figure 1-1. Network Solutions logo, circa 1995. Reproduced with certain permissions from Web.com.

Beyond its important role in the administration of domain names, which helped pave the way for the web's rapid commercialization, Network Solutions also became a particularly noteworthy investment—creating billions of dollars of value during the five short years that we owned the company. As the sole registrar and registry for domain names during the go-go growth years of the Net during the mid-to-late 1990s, the company established new records for growth in share price and market capitalization on Wall Street.

We learned many lessons from our experience with Network Solutions, but five lessons in particular stand out in my mind. Collectively, they are the theme for this book.

First, you've got to be ready to exploit opportunities when they make their way to your front door. This requires designing and implementing a highly flexible and responsive company structure that is quickly scalable to accommodate rapid growth. Technological change is extremely rapid and hard to predict. This was certainly the case for Network Solutions and the fast-changing technology propelling the Internet. During the course of the five-year period that SAIC owned NSI, from 1995 through 2000, the Internet's killer app—the World Wide Web (or just the web for short, developed by Tim Berners-Lee in 1990)—took the world by storm. This technology shift resulted in an immediate and dramatic increase in the demand for domain names, and NSI had to rise to the challenge.

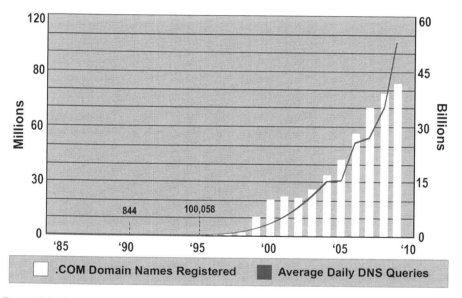

Figure 1-2. Growth of .com domain name registrations and DNS queries, 1985 through 2009. *Source*: VeriSign, January 2010. *Graphic by*: Jamie Dickerson, The Foundation for Enterprise Development.

Second, building a solid financial base is a critical element in any company's long-term success. In my experience, it's better to generate the cash you need for growth by becoming a profitable enterprise than it is to rely on intensive ongoing cash investment from venture capitalists or from the public (via initial public offerings). In the case of Network Solutions, the company was rarely profitable. As a result, its owners were constantly beating the bushes for bank loans and for other sources of the cash they needed to keep the company afloat. By the time we acquired NSI, the company was $7 million in debt, and the owners were under extreme pressure to sell out.

Third, making money out of nothing takes a unique combination of creativity, smarts, guts, and yes, some luck too. Network Solutions was a pioneer in monetizing the web—no other company that I'm aware of aside from a few of the earliest Internet Service Providers (ISPs) had made money from the Internet in the mid-1990s. As such, Network Solutions provided a model for other companies to follow as it tried its own approaches to monetizing the web. It wasn't easy—people weren't happy when NSI started charging a fee for domain name registrations. But having the government continue to pay wasn't an option, and NSI's fee stuck, too.

Fourth, while I have had the good fortune to work with many great people within the United States federal government, if you end up on the bad side of certain politicians, they can make your life a living hell and jeopardize all the good work you have done. Along with NSI, we were constantly assessing the political environment, working hard to anticipate potential political problems and head them off before they emerged. I personally believe that companies and the government should work together instead of in conflict. Our nation's future depends on it.

Fifth and last, but not least, the most effective executives and owners recognize when it's time to sell out and move on to the next opportunity. Nothing lasts forever, and that was definitely the case with the development of the Internet bubble during the late-1990s and its subsequent (and inevitable) bursting in 2000. We sold NSI at the peak of its value; then we used the proceeds in part to fund the pursuit of new acquisitions and other business opportunities. Many of these opportunities continue to pay off for SAIC even today.

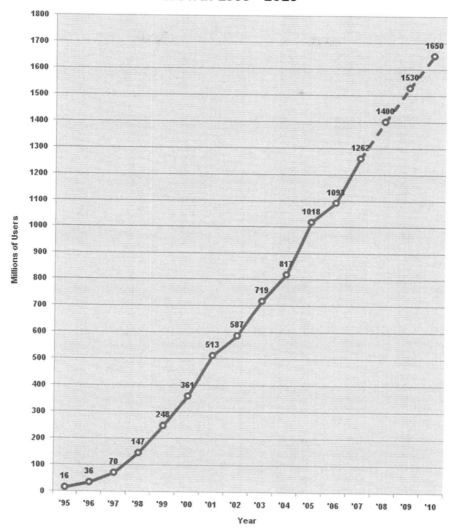

Figure 1-3. Growth of Internet Users, 1995 through 2010. *Source:* www.internetworld-stats.com – January, 2008. Copyright © 2000 – 2012, Miniwatts Marketing Group. All rights reserved.

As you read the chapters that follow, keep these five lessons in mind and consider how they might affect your own business—today and in the future. In the final chapter of this book, I will specifically connect these lessons with the issues facing technology entrepreneurs—and show how the lessons we learned from our experience with Network Solutions can be leveraged to the benefit of your own business.

PART TWO

SETTING THE STAGE FOR EXPONENTIAL GROWTH

A s a company leader, one of your most important duties is to prepare your organization to take advantage of opportunities as they present themselves. While this has always been true throughout the history of business, the ever-increasing speed of business and technology change today requires an even greater attention to this vital task. If your company is not prepared to exploit an opportunity—no matter how small or how large it may be—you risk falling behind the competition and the possibility that the opportunity will never again present itself to you and your people.

When we built SAIC, we created an organization that was uniquely suited to accommodating opportunities for growth—whenever those opportunities arrived, and wherever they arrived from. In my book *The SAIC Solution: How We Built an $8 Billion Employee-Owned Technology Company,* I describe the company's fast-and-flexible organizational structure in the following way:

> I made sure that SAIC's organizational structure remained extremely fluid and flexible, even as the pressures of growth pushed us to make things more rigid and inflexible. If a new opportunity presented itself, we could immediately respond by trying to fit it into an existing unit. If for some reason that approach didn't work, we could create an entirely new business unit. Similarly, an existing unit could just as quickly be dissolved or folded into another unit if a customer's funding evaporated, or if the customer wanted it to be. As the organization grew, SAIC's hierarchy took on a distinctly flat character—when an organizational unit reached a certain size, it would often be divided in two. These two units would grow, and be divided again—creating new operations and groups in the process.[4]

It is incumbent on leaders of businesses in fast-growing industries to prepare their companies for high rates of growth in order to help insure their ultimate success. This means first recognizing that the industry is one currently experiencing, or soon to experience, extreme growth. The company's leadership team must then design an organizational structure that is fast, flexible, and adaptable to rapid changes in the business environment—one where talented managers and employees throughout the organization are willing and able to pursue opportunities as they arise. Finally, sufficient financial resources must be in place to support the company's growth.

In this part, we will take a look at seeds of the Internet industry's growth and the birth and growth of Network Solutions—ending with the event that would change everything, both for the company and for the Internet itself: the award of the National Science Foundation cooperative agreement in January 1993 to serve as the Network Information Services Manager for NSFNET and the National Research and Education Network (NREN) and to provide InterNIC registration services.

2

Understanding the Domain Name System

I love the fact that the DNS works so well in practice, if not in theory.[5]
— Paul Mockapetris, Chairman & Chief Scientist, Nominum

I f you take a fifty-thousand-foot view of the impact of technological innovation on the Internet, you will clearly see that change was driven both from the outside—by the development and implementation of new computer, communications, and information technologies—and from the inside, as a direct result of the rapidly accelerating growth of the Internet. This was clearly the case with the creation of the domain name system.

The domain name system (DNS) is a system for converting host names and domain names into IP addresses on the Internet or on local networks that use the TCP/IP protocol.

When there were only a handful of computers attached to the Internet, there was no need for a domain name system. (The beginnings of the Internet are detailed in Appendix 1, "A Brief Internet Primer [From the Beginning to DARPA].") As the number of computers on the Internet surged upward, however, something had to be done to enable the Internet's continued growth—the original naming system rapidly became inadequate. While the development of the domain name system didn't by itself ensure the ultimate success of the Internet, an effective system was essential to drive the Net's widespread adoption by commercial enterprises, setting the stage for the dot-com boom of the 1990s.

Not only that, but some organization or someone would be needed to administer the DNS—keep it up and running, solve problems as they arose, and make improvements. Thus, after several stops along the way—including SRI International—the NSF established a cooperative agreement with Network Solutions. To better understand exactly what it is that Network Solutions was responsible for under this agreement—and why it was that the company gained such a pivotal role in the transition of the Internet into a vibrant tool for business—let's take a brief look at the evolution of the domain name system.

The Internet (short for "internetwork") is a global network of networks, connected through gateways ("routers") using a common internetworking standard: the Internet Protocol Suite. The Internet Protocol Suite—devised by Vint Cerf and Bob Kahn and described by them in a IEEE paper published in May 1974—is commonly known as TCP/IP for its two most important protocols: Transmission Control Protocol (TCP) and Internet Protocol (IP).[6] The word *Internet* was first used in "Specification of Internet Control Program," written by Vint Cerf, Yogen Dalal, and Carl Sunshine and published as Network Working Group Request for Comments (RFC) 675 in December 1974.[7]

The Internet enables a computer in one location to communicate with any other computer on the network, whether it's across the hall or on the other side of the globe. To accomplish this feat, the Internet must know the location—that is, the specific address—of both source and destination nodes on the network. The Internet Protocol portion of the Internet Protocol Suite provides these addresses, albeit in a way that is not easy for people to recognize or remember. Why? Because it does so in the form of a 32-bit binary number in the case of the original system—Internet Protocol Version 4 (IPv4)—and a 128-bit binary number in the case of the newer Internet Protocol Version 6 (IPv6) system.

IPv4 addresses are most often represented in dot-decimal notation, specifically, four sets of numbers—each ranging from 0 to 255—separated by a dot. So, for example, the IPv4 address for the Google website is 64.233.169.147, while the IPv4 address for the Walt Disney Company website is 199.181.132.250.[8] These fixed addresses enable any computer on the Internet to find its way to the computer services running the Google or Disney websites within a fraction of a second. There's just one problem. While these numerical IP addresses can be handled easily by computers, they are not so easy for humans to work with or to remember. Imagine having to memorize a list of ten different IPv4 numbers just to get to your ten favorite websites. While you might eventually be able to memorize the list, getting to those websites will take a significant amount of time.

But what if there was a better approach—one that works quickly and easily while being friendly to regular, nontechnical Internet users? Actually, there is a better approach, and it is called the domain name system.

The domain name system assigns a series of human-readable letters and/ or numbers to a machine's numerical address on the network (Paul Mockapetris invented the system in 1983 at the request of Jon Postel). The domain name representing Network Solutions' numerical address on the Internet (205.178.187.13), for example, is the familiar networksolutions.com.

A rudimentary version of a domain name system was originally proposed for the ARPANET by Peggy Karp of Mitre in RFC 226 in September 1971 and implemented in October 1971. These "host mnemonics" comprised a simple table of the twenty computer hosts then on the young network (implemented in a computer text file named HOSTS.TXT and first distributed to host operators in 1972).[9]

While this sort of name-address table was fine when there were relatively few hosts on the network, as the host population continued to grow, the table became increasingly unwieldy. David Mills of COMSAT Laboratories in RFC 799 first described the need for an improved naming system. Wrote Mills in September 1981:

In the long run, it will not be practicable for every Internet host to include all Internet hosts in its name-address tables. Even now, with over four hundred names and nicknames in the combined ARPANET-DCNET tables, this has become awkward. Some sort of hierarchical name-space partitioning can easily be devised to deal with this problem; however, it has been wickedly difficult to find one compatible with the known mail systems throughout the community. The one proposed here is the product of several discussions and meetings and is believed both compatible with existing systems and extensible for future systems involving thousands of hosts.[10]

The story of Network Solutions was very much a story about the people who guided the company and who worked for it. The same is true when it comes to the Internet as a whole. And if there's one person who can be credited with having the most influence over the day-to-day workings of the Internet, then that person would arguably be Jon Postel. Jon Postel was the director of the University of Southern California (USC) Information Sciences Institute's (ISI) Computer Networks Division, based in Marina del Rey, California. He led numerous Internet infrastructure activities, was founder and head of the Internet Assigned Numbers Authority (IANA), RFC Editor, and chief administrator of the .US domain. At UCLA, he was involved in the beginnings of ARPANET and the development of the Network Measurement Center, where—working for Internet pioneer Leonard Kleinrock—he helped connect the first computer to the ARPANET.[11]

In a 1997 article, *The Economist* describes Jon Postel in religious terms:

GOD, at least in the West, is often represented as a man with a flowing beard and sandals. Users of the Internet might be forgiven for feeling that nature is imitating art—for if the Net does have a god, he is probably Jon Postel, a man who matches that description to a T. Mr. Postel's claim to cyber-divinity, besides his appearance, is that he is the chairman and, in effect, the sole member of the Internet Assigned Numbers Authority, the organization that coordinates almost all Internet addresses. But unlike God, Mr. Postel's influence, though considerable, does not stretch as far as omnipotence.[12]

In one of his many roles in helping develop and build the ARPANET and, eventually, the Internet, Jon served as ARPANET's Numbers Czar. This involved personally keeping, according to legend, a registry of the ARPANET's IP addresses in a paper notebook.[13] So whenever a new computer needed to be added to the Internet, Postel personally assigned a new IP address to the requester and wrote it down in his notebook, and ultimately in the HOSTS. TXT computer file. Eventually, the fast-growing number of requests for new

IP addresses became too great for one person to handle, so Postel created the Internet Assigned Numbers Authority in the early '70s to take over administration and allocation of IP addresses.[14]

By 1982, the HOSTS.TXT file, which at the time was maintained on a host computer at the SRI International Network Information Center (SRI-NIC), was becoming large and unwieldy as the number of computers attached to the ARPANET grew to more than four hundred. The Network Information Center—run by SRI International from 1970 until 1991—was the information hub first for the ARPANET and later for the Defense Data Network (DDN).[15] In addition, the central nature of its administration and distribution began to conflict with the evolving, distributed model of the Internet. As a result, in 1982 Jon Postel cowrote (with Zaw-Sing Su at SRI) a new Request for Comments—RFC 819—titled "Domain Naming Convention for Internet User Applications." In this RFC, the authors outlined a hierarchical, treelike structure that would enable IP addresses to be distributed across the ARPANET instead of residing only on the SRI-NIC computer.[16]

After publishing RFC 819, Postel asked Paul Mockapetris to help him with the task of developing this new system: the domain name system.

Paul Mockapetris joined USC's Information Sciences Institute in 1978, where he worked with Jon Postel. Mockapetris's response to Postel's request was RFC 882, "Domain Names—Concepts and Facilities," and RFC 883, "Domain Names—Implementation and Specification." In these two RFCs, Mockapetris introduced and outlined a conceptual framework for the domain name system. In a 1988 paper written with Kevin Dunlap of Digital Equipment Corporation, Paul Mockapetris states, "The genesis of the DNS was the observation, circa 1982, that the HOSTS.TXT system for publishing the mapping between host names and addresses was encountering or headed for problems."[17]

In the overview to RFC 883, Mockapetris describes the goal of the domain name system as follows:

> The goal of domain names is to provide a mechanism for naming resources in such a way that the names are usable in different hosts, networks, protocol families, Internets, and administrative organizations.[18]

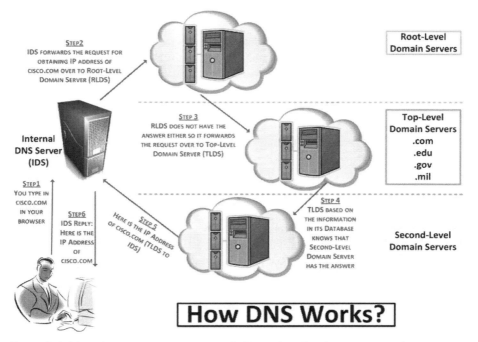

Figure 2-1. How domain name servers work. Reproduced with permission of Najmus Qazi.

Over the years, Paul Mockapetris, Jon Postel, and other computer scientists filled in the blanks that Mockapetris left in his original DNS framework, defining the specific structure of the names in the domain name system and introducing the now-familiar top-level domain names (TLDs) such as .com, .org, .net, .edu, and .gov. In March 1994, Postel summarized the structure of the domain name system as it had evolved up to that time in RFC 1591. Wrote Postel:

> In the Domain Name System (DNS) naming of computers there is a hierarchy of names. The root of system is unnamed. There is a set of what is called "top-level domain names" (TLDs). These are the generic TLDs (EDU, COM, NET, ORG, GOV, MIL, and INT), and the two letter country codes from ISO-3166. It is extremely unlikely that any other TLDs will be created.

Under each TLD may be created a hierarchy of names. Generally, under the generic TLDs the structure is very flat. That is, many organizations are registered directly under the TLD, and any further structure is up to the individual organizations.

In the country TLDs, there is a wide variation in the structure, in some countries the structure is very flat, in others there is substantial structural organization. In some country domains the second levels are generic categories (such as AC, CO, GO, and RE), in others they are based on political geography, and in still others, organization names are listed directly under the country code. The organization for the US country domain is described in RFC 1480.

Each of the generic TLDs was created for a general category of organizations. The country code domains (for example, FR, NL, KR, US) are each organized by an administrator for that country. These administrators may further delegate the management of portions of the naming tree. These administrators are performing a public service on behalf of the Internet community. Descriptions of the generic domains and the US country domain follow.

Of these generic domains, five are international in nature, and two are restricted to use by entities in the United States.

World Wide Generic Domains:

COM: This domain is intended for commercial entities, that is companies. This domain has grown very large and there is concern about the administrative load and system performance if the current growth pattern is continued. Consideration is being taken to subdivide the COM domain and only allow future commercial registrations in the subdomains.

EDU: This domain was originally intended for all educational institutions. Many universities, colleges, schools, educational service organizations, and educational consortia have registered here. More recently a decision has been taken to limit further registrations to four-year colleges and universities. Schools and two-year colleges will be registered in the country domains (see US Domain, especially K12 and CC, below).

NET: This domain is intended to hold only the computers of network providers, that is the NIC and NOC computers, the administrative computers, and the network node computers. The customers of the network provider would have domain names of their own (not in the NET TLD).

ORG: This domain is intended as the miscellaneous TLD for organizations that didn't fit anywhere else. Some non-government organizations may fit here.

INT: This domain is for organizations established by international treaties or international databases.

United States Only Generic Domains:

GOV: This domain was originally intended for any kind of government office or agency. More recently a decision was taken to register only agencies of the U.S. federal government in this domain. State and local agencies are registered in the country domains (see U.S. Domain, below).

MIL: This domain is used by the U.S. military.

Example country code Domain:

U.S.: As an example of a country domain, the U.S. domain provides for the registration of all kinds of entities in the United States on the basis of political geography, that is, a hierarchy of <entity-name>.<locality>.<state-code>.US. For example, "IBM.Armonk.NY.US". In addition, branches of the U.S. domain are provided within each state for schools (K12), community colleges (CC), technical schools (TEC), state government agencies (STATE), councils of governments (COG), libraries (LIB), museums (MUS), and several other generic types of entities (see RFC 1480 for details).[19]

By early 1985, the domain name system was up and running, and on March 15, 1985, the very first domain name—symbolics.com—was registered through the new DNS process. This initial registration was soon followed by bbn.com on April 24, 1985; think.com on May 24, 1985; mcc.com on July 11, 1985; dec.com on September 30, 1985; and northrop.com on November 7, 1985.[20]

Reflecting back on his role in the creation and evolution of the domain name system in a 2003 interview, Paul Mockapetris had this to say:

I designed the technology to allow you to do domain names, but I didn't define which domain names we would use. For example, the people in the UK prefer .ac (academic community) as opposed to .edu. I had made none of those choices, because I didn't want to get involved in those politics. Then when it was time to define them, there were all sorts of bad ideas about how to select domain names. So that, for example, there was one famous individual who said, "In all domain names, the most important part should be a country code," instead of having .com and so forth. I've always been in the mode of thinking that experimenting when you can afford it is a much better idea. So I said—well let's try both and see what works. So the only reason we have .com today is that finally people got tired of arguing with me about that and said, "Well we'll try it, but no one will ever use it." If they had known what was going to happen they would never have agreed! It wasn't that I thought the .com was the right idea, I thought that trying both country codes and generic and see what worked was the right thing. I tend to think that that kind of experimentation was a good idea, and I wish we had done more of it![21]

Domain names—and their mapping to specific IP addresses—are propagated across the entire Internet by root name servers, computers that reliably publish the contents of the root zone file. The root zone file contains a list of names and numeric IP addresses of the authoritative DNS servers for all top-level domains (such as .com, .org, and .edu), and the country code top-level domains.[22]

There are thirteen different root name servers. Each is indicated by a letter, A through M, and each is represented by actual physical servers located at one or more sites scattered around the world. The thirteen root server operators are:

A: VeriSign Global Registry Services (note: this is the A server that used to be operated by Network Solutions and that figures prominently in Chapter 11: The Night the Internet Died)

B: Information Sciences Institute

C: Cogent Communications

D: University of Maryland

E: NASA Ames Research Center

F: Internet Systems Consortium, Inc.

G: U.S. DOD Network Information Center

H: US Army Research Lab

I: Autonomica/NORDUnet

J: VeriSign Global Registry Services

K: RIPE NCC

L: ICANN

M: WIDE Project[23]

As noted above, more than one actual server physically supports many of these lettered root name servers. For example, VeriSign's A server is currently represented by servers at six different sites, including Los Angeles; New York City; Palo Alto; Ashburn, Virginia; Hong Kong; and Frankfurt, Germany. VeriSign's J server is currently represented by servers located at seventy different sites, including Prague, Tokyo, Kuala Lumpur, Cairo, Moscow, Rome, Paris, and San Francisco.[24] The operators of these root name servers do not create, nor do they edit, the root zone files—this is the job of IANA. Instead, it is the job of the root zone file operators to simply publish the files they receive from IANA, exactly as they are received.[25]

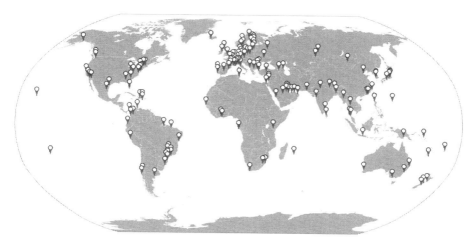

Figure 2-2. Map of root server locations worldwide. *Source*: Root Server Technical Operations Association. *Graphic by:* Jamie Dickerson, The Foundation for Enterprise Development.

The original design of the root server system was limited by the maximum size of the data packet, and the original design only allowed a maximum of thirteen distinct servers that could serve up the root zone—the very root of the tree. When a DNS server starts up, it actually consults the root server—in essence, saying, "Tell me where everything is."

There weren't always thirteen root servers. According to Mark Kosters, the shift from nine servers to thirteen was made possible when Paul Vixie performed some DNS magic. Says Kosters, "I, along with Bill Manning and Jon Postel, renamed the roots from their origin-specific names to root-servers.net. Once done, we realized we could add four more. Postel and I negotiated for him to take two and for NSI to take two. We did this so that we were equal and no one had a majority. The idea was to have them operated by foreign organizations. That happened with two of the four. We gave the first one, K, to RIPE, and Postel tried to give the second one, J, to Japan. I didn't think that was a good idea and that is how they got M. ICANN somehow hijacked L and J stayed at NSI."[26]

Originally, the Defense Advanced Research Projects Agency (DARPA) administered the DNS, assigning domain names to users as necessary. This task was carried out by a DARPA contractor, IANA (the Internet Assigned Numbers Authority), which was run by Jon Postel for more than thirty years. In the late 1980s—when NSF started popularizing the use of the Internet more broadly, well beyond DARPA's contract awardees—DARPA suggested that the National Science Foundation take on the duties of administering the non-military portions of the DNS and IP allocations. NSF agreed to do so.[27]

Congress created the National Science Foundation in 1950 "to promote the progress of science; to advance the national health, prosperity, and welfare; to secure the national defense." Today, the agency has an annual budget of about $6.9 billion, and it funds approximately 20 percent of all federally supported basic research conducted by America's colleges and universities via more than ten thousand new limited-term grant awards each year. NSF is the largest source of federal research dollars in a number of fields, including computer science, the social sciences, and mathematics.[28]

It was the Coordinated Experimental Research Program within the Computer Science Section of the agency's Mathematical and Physical Sciences Directorate that in 1981 created NSF's first computer network, dubbed CSNET (for Computer Science Network). This early network—connected to ARPANET, which at first used the Network Communication Program (NCP) interface, with plans to move to TCP/IP—provided e-mail and general Internet services to its users. In addition, it opened the first Internet gateways between the United States and countries in Asia and Europe.

NSF charged its users a fee to access CSNET—starting at $2,000 for small computer science departments and running up to $30,000 for large industrial users. By 1986, the network had more than 165 users—including university and government computer research and industrial organizations—and it was financially self-supporting. In 1984, NSF launched its most ambitious computer project yet: the supercomputing program. In 1985, NSF funded the establishment of four supercomputer centers at the University of California, San Diego; the University of Illinois; Princeton University; and Cornell University. A fifth supercomputer center was later established in Pittsburgh, jointly run by the University of Pittsburgh, Carnegie-Mellon University, and Westinghouse.

One of the goals of the establishment of these supercomputer centers was to create a new high-speed computer network that would become "a network of networks"—an Internet—and dramatically increase the speed of data transmission, hopefully up to twenty-five times faster than CSNET. This network of networks came to fruition in 1986 when the five supercomputer centers were connected—creating the NSFNET, which also connected to ARPANET.[29] In 1990, IBM, Merit, and MCI teamed up to form the nonprofit Advanced Network and Services, Inc. (ANS), to operate NSFNET.[30]

Demands on the network continued to grow. NSFNET was significantly upgraded in 1988 to create an Internet backbone to support and bring together a mishmash of existing networks, including thirteen different regional networks and supercomputer centers comprising more than 170 different university campus computer networks. This T-1 network operated at 1.5 megabits per second (Mb/s)—significantly faster than ARPANET—and transmitted more than 152 million packets of information a month, with usage increasing at a rate of 10 percent each month. The network quickly went to 45 Mb/s.[31]

In an interview for this book, George Strawn—director of the National Coordination Office for Networking and Information Technology Research and Development (NCO/NITRD) and NSFNET program director from 1991 through 1993—explained his agency's philosophy when it agreed to take on administration of the DNS: "We agreed that we would run a solicitation and become responsible for supporting the domain name registration. This was primarily financial support—we didn't really expect to take over or change any of the technical management aspects that DARPA already had in place with the IANA or anyone else."[32]

In March 1991, something happened that many consider the birth of the commercial Internet that we know today: the NSFNET acceptable use policy—originally created in 1989—was altered to allow commercial traffic. The original general principle of the NSFNET backbone services acceptable use policy limited use of the NSFNET by commercial organizations to narrow circumstances:

GENERAL PRINCIPLE:

(1) NSFNET Backbone services are provided to support open research and education in and among US research and instructional institutions, plus research arms of for-profit firms when engaged in open scholarly communication and research. Use for other purposes is not acceptable.[33]

In addition, the acceptable use policy clearly spelled out unacceptable uses, including those uses carried out for commercial purposes:

UNACCEPTABLE USES:

(10) Use for for-profit activities unless covered by the General Principle or as a specifically acceptable use.[34]

Before this change was made, with few exceptions, use was limited to people and institutions doing scientific research or working on government contracts. One such exception was the permission granted to Vint Cerf in 1988 to connect MCI Mail to NSFNET, which Cerf accomplished in 1989.[35] So, when the doors of NSFNET were thrown open to the world at large, usage climbed steeply. In 1991, NSFNET linked together 16 different sites and more than 3,500 different networks. Just one year later—in 1992—the number of connected networks had mushroomed to more than 6,000, with more than 2,000 of those located outside of the United States. And while NSFNET averaged 1.32 trillion bytes of information transmitted each month in March 1991, by December 31, 1994, this number had shot up to an average of 17.8 trillion bytes per month.[36]

There is one more piece of the going-commercial story to be considered. For commercial businesses to offer commercial packet-switched service, the FCC had to first approve it. The reason is that the Communications Act of 1934 (as amended) rate-regulated common carriers. Users of common carrier services dealt directly with licensed carriers. There were no middlemen allowed. That is, a subscriber could lease capacity from AT&T, for example, for personal or organizational use but could not resell it to the public. That would make that subscriber a common carrier, in violation of the act. This, in effect, prohibited the Internet Service Provider business. The FCC approved packet-switched service sometime in the 1980s, with ISPs buying capacity wholesale from carriers such as AT&T and then reselling it at the retail level to Internet subscribers.[37]

According to Vint Cerf, the National Science Foundation played a key role in opening the Internet for business. Says Cerf, "There were some decisions made at the NSF at certain points which I think are very, very relevant to the commercialization of the Internet. Clearly the first one was [when Vint Cerf was given] permission to connect a commercial system up to the NSFNET backbone. That essentially broke the logjam that had prevented a commercial engine from starting and helping to support and propagate the network. There was legislation in 1992 that specifically permitted commercial traffic to flow on the Internet backbone. My recollection is that then-Vice President Al Gore was very instrumental in getting the legislation passed."[38]

The commercial system Cerf refers to was MCI Mail—the first commercial Internet electronic mail service—which NSF gave him permission to connect to NSFNET in 1988, breaking the Appropriate Use Policy. And the legislation that he refers to is the Scientific and Advanced-Technology Act of 1992, sponsored in the Senate by Senator Barbara Mikulski and introduced in the House of Representatives by Congressman Rick Boucher, which became Public Law No: 102-476 on October 23, 1992.[39] This new law amended the National Science Foundation Act of 1950 to authorize the NSF to "foster and support access by the research and education communities to computer networks which may be used substantially for purposes in addition to research and education in the sciences and engineering, if the additional uses will tend to increase the overall capabilities of the networks to support such research and education activities."[40]

Internet pioneer Bob Kahn also believes that the NSF played a vital role in expanding the role of the Internet. Says Kahn, "When NSF jumped in, they broadened it up much more significantly to virtually the entire science and education community in the country that they dealt with, and with some focus on international connectivity to the NSFNET—first, I think, in Europe and then eventually to Japan and elsewhere. So I think NSF really gets a lot of the credit for opening things up. DARPA kept the nascent Internet fairly closely held. If you wanted to get on it in DARPA days, you probably had to be a DARPA contractor and get their approval. NSF took the initiative with the NSFNET to open it up. They funded regional nets of different sorts. They empowered those folks to make their own deals, and get their customers on as they saw fit. I think by the mid 1990s, that strategy had taken hold and the Internet started to grow significantly."[41]

As Internet traffic continued to climb, and as commercial interests became increasingly engaged in exploring the many business possibilities of the Internet, the National Science Foundation began to look for ways to get out of the business of running the NSFNET without harming the system. And the private sector also began to push for NSF to get out of the business of running the Internet. Bill Schrader, cofounder of the one of the first commercial Internet service providers (ISP)—PSINet—called for hearings, which the House Committee on Science held later in that year. His goal (shared by others who were leading the charge for commercialization of the Internet) was to get the government out of the way so the private sector could take over the Internet.[42]

Another issue that arose while NSF was architecting its approach to accommodating the private sector was whether the agency should retain just the one backbone that it already had or build a second Internet backbone for competition's sake. George Strawn canvassed the regional networks and their members and found out that there was already great consternation in the community about the NSF networks. The private-sector members in particular were chomping at the bit and felt something had to change because NSFNET had an unfair competitive advantage due to the government subsidies that supported its existence. According to George, after gathering input, he told his boss, "I've got a new answer for you. The answer is we should have neither one nor two backbones—we should have zero backbones and turn it entirely over to the private sector. That played very well with our regional net constituency and seemed to play well with everybody else. And that's what we did—we got out of the backbone business."[43]

For a government employee, removing the vital Internet backbone functions from direct government control and entrusting them to the private sector was a risky move. If anything went wrong in the transition, or if the private sector dropped the ball, then George Strawn and the top brass at NSF would find themselves in hot water. But the government and the involved private-sector companies were committed to working together instead of against each other. And as a result the growth of the Internet was able to continue unimpeded.

On April 30, 1995, the National Science Foundation decommissioned NSF-NET, and its backbone functions were privatized by a combination of PSINet (which had first been the New York State Regional Network backbone before it went national), MCI (which had been the backbone provider for NSFNET), Sprint (which had been the provider for international connection services, to connect foreign institutions of higher learning to the NSFNet), and UUNET (which had been a DARPA contractor for some years, well before the birth of the Internet).[44] And, for NSF, the timing could not have been better. Within a couple of months after the National Science Foundation shuttered NSFNET, Internet traffic went exponential—driving these four companies to their knees as they tried to keep up with the resulting massive increases in business.

The National Science Foundation wasn't entirely out of the Internet business yet—it still had the primary responsibility for administering the DNS. And little did George Strawn or anyone else at NSF know, that little corner of the Internet was just about to go exponential, too.

In 1990—as NSF began to grapple with the idea of taking NSFNET out of service—English engineer and computer scientist Sir Tim Berners-Lee, a researcher at CERN, the European Laboratory for Nuclear Research, developed and implemented a new Internet-based innovation: the World Wide Web.[45] Using the web, users could for the first time easily embed Internet addresses into their documents using hyperlinks, which people could click to be connected to the referenced location—a location that could be anywhere on the Net.

Again, Internet technology was moving fast, but this innovation did not go unnoticed. A programming team at the National Center for Supercomputing Applications (NCSA) led by Marc Andreessen at the University of Illinois developed a graphical web browser—given the name Mosaic—that allowed people to post web pages that contained different sizes, colors, and styles of text, images, sound files, video, and more. In 1994, Andreessen left the NCSA and cofounded Mosaic Communications Corporation (later renamed

Netscape Communications) with Silicon Valley entrepreneur Jim Clark. By 1996, the Netscape browser dominated the web, with more than 70 percent of web users using it.[46]

In the spring of 1995, George Strawn received an offer to become NSF's new division director for networking. He accepted the position, and the first decision awaiting his arrival was the domain name management issue. Requests for domain names were going through the roof, and based on their back-of-the-envelope estimates, the cost of administering the DNS would likely explode from the then-current budget of about a million dollars a year to something more on the order of a million dollars a month. Fortunately for NSF, the agency had in 1993 entered into a cooperative agreement with a small business in Herndon, Virginia, by the name of Network Solutions, Inc., to run the Internet's domain name registration service. It would be the responsibility of NSI and its small staff to deal with the coming tidal wave of demand for domain names.

3

A Company by the Name of Network Solutions

NSI's principals shared a vision of "distributed processing" as the wave of the future. They believed technology would shift from mainframe computers to distributed microprocessors...they were right. However, that fact did not make it any easier getting NSI off the ground.

— Cliff Hocker, *Black Enterprise* magazine[47]

The history of business is chock full of examples of companies that achieved great things—pioneering new markets where there were none, generating substantial profits for owners and investors, and making a positive and lasting change in the world. But what is it that makes these companies great? In many cases, it's a matter of having the right combination of inspired and talented people with a strong vision of the future, at the right time and place, with the right offering of products or services. Achieving this happy state, however, often means anticipating and then staying atop a wave of change in the marketplace while others either don't notice that a wave is coming or wait to react until it has already passed.

In the case of Network Solutions, the company's key people anticipated the shift from standalone mainframe computers to distributed networks of mini and personal computers. But while Network Solutions found significant success in the computer networking market, it would be some years after the company's founding before the next wave of change would arrive, and NSI's ultimate destiny would be revealed.

The story of Network Solutions, Inc., begins with Gary Desler, who founded the company in 1975 as a sole proprietorship.[48] Born and raised in

Nebraska, Gary joined the navy in 1960 to get electronics and computer training. During the course of his enlistment, he spent two years on a diesel submarine and five years on a nuclear submarine (the second Polaris missile-firing submarine, *Patrick Henry*). He left the navy in 1968, attended the University of Maryland, and then went to work for Digital Equipment Corporation (DEC), where he worked on its PDP-10 mainframe computer product line.

At the time (late 1960s/early 1970s), the PDP-10 was the largest time-sharing computer, and it cost about one-tenth the price of the IBM equivalent. It was mainly for these reasons that when the ARPANET was established, most of the computers on it were PDP-10s—often running BBN's Tenex operating system. While working for DEC, Gary gained extensive experience in the area of computer networking—primarily building networks for the Department of Labor—and became one of the company's leaders in the field. With a growing family, and no raises because of Nixon-era wage and price controls, he made the move to the federal government, accepting a job with the US Public Health Service in Washington, DC.[49]

While Gary's move to the Public Health Service provided him with a much-needed boost in income, finances were still tight. So in 1975, he decided to moonlight, starting his own business—a sole proprietorship—in his free time.[50]

The name of this business? Network Solutions.

In the company's earliest days, the focus of Network Solutions was the installation of inexpensive line drivers in place of the expensive Bell modems that companies up to that time were required to install on each end of individual wire runs (owned by the phone company) within a building or other facility. Network Solutions would buy the new line drivers for just a couple hundred dollars each and run new copper twisted pair wires that the customer would own from office to office. In this way, Gary and his new company could replace $5,000 a month in recurring phone company communication charges with a one-time fee built from just $400 in labor and $400 worth of parts. This made Network Solutions increasingly popular with a growing roster of clients.

In about 1977, Gary formed the Washington Area Teleprocessing Users Group (WATUG) with Ed Peters and Howard Berkowitz from the Department of Labor. Gary had met Ed when one of his clients at Digital Equipment

Corporation was the Department of Labor. The group attracted all of the big timesharing users in the DC area—about thirty people—who would meet to discuss common technical problems and strategies for dealing with the phone company. The more Gary and Ed worked together, the more they realized there was a major business opportunity for a company that specialized in helping businesses with their timesharing problems.[51]

In the summer of 1978, Gary and Ed decided that the time had come to start up their new business. When Ed Peters's college roommate from his days at Catholic University found out that Gary and Ed were talking about forming a company in the networking arena, he told them that he knew two men who also wanted to start something but didn't have any business plan. The two men were Emmit McHenry and Tyrone (Ty) Grigsby.[52]

Ty Grigsby earned a BA in business administration from Howard University in 1968, and he graduated from the American Management Association's (AMA) Management Internship Program in 1969. His career began in 1969 as an investment analyst with The Travelers Insurance Company in Hartford, Connecticut.[53] Says Grigsby about his career choice, "I had finished the AMA graduate program and went to work with Traveler's in 1969—it was about a year after Dr. King had been killed, and there were a lot of riots and turmoil in the inner cities. During that time, many of the insurance companies were bringing in a lot of black executives in entry-level positions. And that's where I met Emmit. We were part of Ujamaa, which was a group that met with chief executives of corporations, city council people, mayors, and other people in positions of power."[54]

In 1973, Ty decided he had had enough of New England and moved back to Washington, DC, to join Raven Systems and Research, a computer services company started by his friend Ray Mott, where he took the position of executive vice president. Raven had won a contract to run the executive office of the president's computer center. Ray had no operations or marketing experience, so Ty was brought in to handle the administrative functions of the business, including legal, financial, administrative, and marketing duties. After five years with the company, Ty left Raven Systems and Research in November 1978 and decided to get in touch with his friend Emmit McHenry to discuss the idea of starting a company of their own.[55]

Emmit was born in Forrest City, Arkansas, and grew up in Tulsa, Oklahoma. He earned his BA in communications at the University of Denver in 1966 and was then drafted into military service. After two and a half years in the US Marine Corps, Emmit left the military and enrolled in graduate school at Northwestern University, where he earned his master's degree. Emmit went on to become a systems engineer with IBM and regional vice president for Allstate Insurance Company, where he was responsible for the company's operations in the Pacific Northwest, Alaska, and Hawaii.[56]

After Ty and Emmit discussed the idea of starting their own company and decided to move forward, Ty contacted another friend—Sam Harrison, an executive search consultant who he had met during his Raven days. According to Ty, it was Sam who introduced him and Emmit to their future business partners. Says Ty, "I told Sam what I was trying to do, and he identified Gary Desler and Ed Peters. He set up a meeting for all four of us at his office on Connecticut Avenue in Washington, DC. We talked about what each of us brought to the party and decided to move ahead."[57]

In January 1979, the four men incorporated their new business—Network Solutions, Incorporated—in Washington, DC. In February 1979, Gary Desler quit his job with the Public Health Service and focused on doing the company's marketing. Ty came on board as the company's CEO, and Ed—who had moved from the Department of Labor to the Defense Communications Agency—provided technical project management skills. Emmit brought strong corporate management and leadership skills, as well as technical expertise gleaned from his time with IBM,[58] and he became the chairman of NSI's board of directors.[59] Emmit didn't immediately join the company, instead keeping his job with Allstate and flying in from Chicago for a day or two at a time for meetings. When he did come on board in 1986, Emmit took over as CEO (retaining his position as board chairman), and Ty retained the position of president and a seat on the company's board of directors.[60]

In March 1979, Network Solutions, Inc., officially opened its doors for business. In the beginning, the company did business in a small office located at 8150 Leesburg Pike, Suite 600, in Vienna, Virginia. This first, one-room office had just three desks in it. The founders specifically chose this area of northern Virginia (coincidentally, just down the street from SAIC's Tysons

Corner offices) in which to locate their business because the area was fast becoming a hotbed for computer technology—several other tech companies were also locating their offices there at the time. The hope was that this location would make it easier to lure talented technical employees.[61] This did indeed turn out to be the case. The company was started on a shoestring, and securing startup financing was a significant and ongoing challenge for the four company founders.

Says Desler, "When we started Network Solutions, Inc., none of us had a dime. We started it with twenty-one Visa credit cards. We took our government retirement money and put it into the business. We sold our houses and lived on that money for a while until things turned around. This was back in the good old Jimmy Carter administration days, and none of us had any money, but we knew what we wanted to do."[62] According to Grigsby, he personally did whatever he could to raise his financial stake for the new venture: "I borrowed money in a variety of ways. I used credit cards, loans from friends, a second-mortgage loan, and bank loans."[63]

While he was at Raven Systems and Research, Ty Grigsby had worked with Paul Brammel, who was president of the National Bank of Washington, a small bank in Washington, DC. Paul had helped Raven obtain a $500,000 line of credit to get work started on a contract the company had won. When Ty and his partners were originally exploring the idea of starting up Network Solutions, Inc., Ty contacted Paul to give him a heads up that there was a good chance he would be contacting Paul in the near future to secure financing for the new business. According to Ty, "Paul said, 'I'll help you in any way I can—come see me when you get started. But don't wait until you need the money.'"[64]

So in April 1979—soon after the partners got NSI off the ground—Ty went to Paul and asked for a loan. True to his word, Paul escorted Ty to one of the bank's loan officers, vouched for him and Network Solutions, and told the loan officer to take care of Ty. Five minutes later, the loan officer approved Network Solutions' first bank loan, in the amount of $10,000. While it wasn't much, combined with their own funds and the help of friends and family, it was enough to sustain the company as the partners turned up the heat on their marketing efforts.

For the first nine months after startup, NSI's owners didn't draw a salary. To celebrate Christmas 1979, however, the four founders agreed to gift one another $2,000 in cash from the company's growing bank account.[65] By 1980—within a year after its incorporation—Network Solutions had ramped up to eleven employees and $300,000 in sales.[66]

One thing that helped NSI build its business was when the company obtained its 8(a) certification. The 8(a) Program (begun in 1968 under the authority of Section 8[a] of the Small Business Act, as amended) is administered by the Small Business Administration (SBA). The program allows federal government agencies to enter into contracts for supplies and services with SBA for performance by businesses in the SBA Business Development Program. SBA then subcontracts the actual performance of the work to small businesses owned and controlled by socially and economically disadvantaged individuals—giving these businesses a chance to win contracts that they might not normally be considered for.[67]

Network Solutions worked with a DC-based attorney who knew his way around federal government contracting circles to put together their application for 8(a) certification, and the application was submitted to the Small Business Administration in the summer of 1979 and approved in February 1980. This approval for inclusion in the federal government's minority set-aside program would turn out to be a critical tool in the company's marketing—and in its eventual growth.

But while this particular obstacle had been conquered, yet another loomed over the company. NSI desperately needed cash to fund its operations, and by the summer of 1979, the founders' initial contributions of capital to the business—and the $10,000 they had received via Paul Brammel—were quickly running out. Without additional capital to keep the company afloat, NSI would soon be out of business. Each of the four founders had business connections they hoped to parlay into an infusion of cash into the business, and they began to push them hard. Again, Ty came through with another connection, Harry Fisher, who was head of the commercial lending department at United Bank. Harry was a seasoned banker and a bit of a risk taker. And he gave Network Solutions its second loan—this one in the amount of $25,000.[68]

Business began to flow into the company—first small-dollar, low-risk projects, but eventually, larger, higher-risk projects. Once the ice was broken, the company was on its way—by 1980 building up a backlog of about $250,000 worth of contracts. This was still not enough revenue to pull the company's bottom line out of the red and into the black, however. NSI was technically bankrupt, with a $90,000 negative net worth, most of it owed to the federal government in the form of payroll taxes. And the IRS wanted its money—sending an account executive to NSI's offices every week seeking payment. Things were so bad for a while that, according to Ty Grigsby, "I fully expected to come there one day and find a padlock on the door."[69]

To shore up its dangerously weak financial position, Network Solutions sent out financing proposals to at least a half-dozen banks. Each of these banks turned down the opportunity to loan money to NSI. That is, until Gary Desler came into the office one day in February 1981 and announced that the day before, while he was at a dentist appointment, the topic of NSI's financial situation came up with his dentist during the course of the exam. Gary's dentist recommended that NSI talk to a loan officer he knew. As it turned out, the "loan officer" was actually the bank's senior vice president of commercial lending for the entire region.[70]

Network Solutions contacted the bank to pursue a loan, with Gary's dentist also calling to vouch for the company. The senior vice president agreed to check out NSI, so he sent out one of his people—Joe Pipitone, a vice president at the bank—to have a look. Joe liked what he saw, and by April 1981, NSI had its third loan—this one a credit line in the amount of $150,000 (an amount that would grow to $6 million over the next five years).[71]

By the end of 1982, NSI's annual revenues broke the $1 million mark, finishing at $1.1 million. With NSI pulled from the brink of bankruptcy and revenues continuing to grow, the company was finally in the position to turn its attention away from fending off its creditors and the IRS and toward delivering products and services to its customers.[72]

4

GAINING EXPERIENCE WITH THE INTERNET—AND GROWING FAST

I am not surprised that the Internet provides anyone the ability to connect from any location at any time with any device and is invisible. I am surprised at how far the Internet has penetrated every fabric of our lives and society.[73]

— Leonard Kleinrock,
Distinguished Professor of Computer Science, UCLA

While at the Defense Communications Agency, Ed Peters worked on AUTODIN and AUTODIN II, which were data communications services within the Department of Defense. But just as the ARPANET was beginning to gain traction and widespread usage within the Department of Defense, the AUTODIN II program was killed, and the ARPANET became the backbone of the Defense Data Network (DDN). On January 1, 1983, TCP/IP became the standard for the ARPANET.

Soon after the adoption of TCP/IP as the ARPANET standard, the Department of Defense mandated that any computer system sold to the DOD had to have TCP/IP. IBM Corporation did not have TCP/IP, which was a big problem for the company. The immediate solution for IBM was licensing Network Solutions' software on computers to be sold to the Department of Defense. During the course of the 1980s, most of NSI's work was service contracts for the US government. But one contract in particular became especially important for the future of the company. Soon after MILNET was created in 1983—when the military sites split off from the ARPANET, leaving behind the university and other users—NSI won the contract to help administer this new network. Network Solutions branded its TCP/IP software "OPEN-Link" and helped develop some of the early Internet protocols, including the domain name system.

NSI continued to seek business in nondefense federal agencies. Around 1983, the Department of Labor awarded a systems development and network management contract to Network Solutions. Over time, the initial award of $100,000 grew to more than $10 million, greatly contributing to NSI's growth.[74]

In 1983, NSI's four principals created a three-year plan to grow the business through 1984 and 1986. The overall goal of the plan was to firmly establish Network Solutions as "a viable company with long-term growth potential and with a variety of options for converting the equity to liquid assets." It was anticipated that approximately half of this growth would come from the company's existing contract base, while the other half would derive from closely related extensions of this business.[75] In 1984, Network Solutions more than doubled its sales to $7.7 million from $3.3 million in 1983.[76]

In 1985, Network Solutions reached a key milestone along its growth path when IBM awarded the company a large subcontract to provide systems programming, capacity planning, and operating system maintenance on the institutional mainframe computers at NASA's Johnson Space Center in Houston, Texas. Although the incumbent contractor, Computer Sciences Corporation (CSC), was deeply embedded in the work—and in the Space Center—IBM thought it would be worth its while to take a shot at unseating it. To help bolster its proposal, IBM invited Network Solutions to team.

In the world of government contracting, large prime contractors such as IBM are motivated to seek out qualified minority- and woman-owned businesses to team up with. Why? Because the government often awards extra points in its proposal evaluation criteria for proposals that include such "disadvantaged" (but otherwise qualified) businesses.

Ultimately, three companies bid on the NASA contract: IBM, CSC, and Planning Research Corporation (PRC).[77] As it turned out, the IBM/NSI team offered neither the lowest price (it was priced between CSC and PRC) nor the highest technical score (again, it was scored between CSC and PRC), but it did offer the best combination of price and technical capability of the three proposals. As a result, the IBM/NSI team was selected for award by the Johnson Space Center—pushing CSC out of its entrenched position.

While Network Solutions performed its services for the Defense Data Network job, the company was also working with the National Science Foundation (NSF), which had in 1985 established the NSF Network (NSFNET). The company was very much involved in the increasingly rapid evolution of the Internet, even cosponsoring the first Interop conference in Monterey, California, kicking in $5,000 along with two or three other companies. The reason for the conference was to provide engineers with a forum to plug their data WAN (wide-area network) gear together and establish interoperability. According to Gary Desler, "It was a bunch of propeller-heads talking about bits and bytes and mail and file transfer protocol. We had all these dissimilar computers—Burroughs, Univac, Honeywell, IBM, and DEC—and they needed to be interoperable. We worked through all kinds of issues to get the computers to talk with one another over networks. Interop turned out to be huge, and we funded and attended all of the early shows."[78]

In 1986, the company's annual revenues hit the $18.5 million mark—very close to the company's goal of $20 million from the three-year plan NSI's founders had created in 1983. NSI was clearly a going concern, and it was at this point that Emmit McHenry left his insurance company career behind and became a full-time employee of Network Solutions. When Network Solutions was incorporated, the original plan was for Emmit to take the company's position of chief executive officer, with Ty as chief operating officer, Gary as NSI's vice president of marketing, and Ed as vice president of operations. Because Emmit didn't initially join the company, Ty became NSI's CEO, and he remained in this position until Emmit joined in 1986. With Emmit in as CEO, Ty turned his focus to running special projects and troubleshooting projects that were running into problems.[79]

In 1987, Ty took a sabbatical and stepped out of the company's day-to-day operations for six months. During this time, the three remaining principals of NSI decided to move the company to a much larger space at 505 Huntmar Park Drive in Herndon, Virginia.[80] Despite its ongoing financial difficulties, by the early 1990s NSI had grown into a vibrant technology company with about five hundred employees at its peak and a diverse portfolio of government and commercial contracts. Network Solutions was different from much of the competition because it had a strong contract base in the government and private-sector markets. Companies that do business primarily in the government market often have a hard time selling themselves to commercial firms, and businesses that specialize in doing commercial work often find it difficult to make the leap to government contracts. The two worlds are quite different, and different rules apply.

Network Solutions won a four-year, $12 million subcontract with AT&T to provide nationwide network engineering and field service support for the federal government's FTS2000 long-distance telecommunications system. Network Solutions won the subcontract—awarded in August 1990—over a crowded field of seventy other bidders. Other clients included General Electric, Pacific Bell, Commonwealth Edison, the Sacramento Public School System, and the US Department of Health and Human Services.[81]

The Stanford Research Institute (later renamed SRI International) was the first central authority responsible for coordinating the operation of ARPA-NET. Hosts on the ARPANET were given names, and SRI International was responsible for distributing lists of these host names to all interested parties. When the domain name system was created in response to the increasing complexity of the network as the user community grew, domain name registration services were provided by SRI International by way of its Defense Data Network-Network Information Center (DDN-NIC).[82]

In 1991, Network Solutions got a glimpse of its future when Network Information Center (NIC) services were transferred from SRI International in Menlo Park, California, to Government Systems, Inc.—a Chantilly, Virginia-based provider of international communications and network-related services to the US government and other organizations—on October 1, 1991.[83] Soon after SRI International transferred the responsibility for providing DDN-NIC services to Government Systems, Inc., the company subcontracted the work to Network Solutions.

According to the Network Working Group Request for Comments (RFC) 1261, GSI's (and thus NSI's) scope of work included:

> …the transition of services currently offered to DDN and Internet users by
> SRI. These services include network/user registration (i.e., network number

and top level domain name assignment), online information services, and Help Desk operations. GSI will also continue RFC and Internet-Draft archive and distribution services.[84]

Mark Kosters—chief technology officer of the American Registry for Internet Numbers (ARIN), the organization that distributes Internet number resources, including IP addresses, in Canada, many Caribbean and North Atlantic islands, and the United States—joined Network Solutions as a senior engineer in July 1991, after earning his masters in computer science at George Mason University in Fairfax, Virginia.[85] He recalls how busy the company and its employees were at the time. Says Kosters, "At that point Network Solutions had just been awarded the subcontract to run the Defense Data Network-Network Information Center. A part of that included some work funded by the National Science Foundation that did the nonmilitary portions of it—registrations in .com, .net, .gov, .edu, and so on. They were all intertwined when they came over from SRI-NIC." This intertwining of military and nonmilitary Internet functions was becoming a problem as the system became increasingly internationalized and researchers and other Internet users grew wary of the US military's pervasive control of the infrastructure that ran the Net.

Continues Kosters, "I had a forty-hour workweek the first week I was there, and from there on out, it was eighty hours or more every week. At one point I think I tagged out about 130 hours in a particular week, just trying to get through the mountain of work that we had to do. It was a huge process bringing the DOD NIC to operational status, and people didn't think we could do it. We did do it, but it took lots and lots of hours."[86]

The heavy workload that Kosters and the Network Solutions team put in during this period of time was a direct response to the growing popularity of the Internet. But the company and its employees hadn't seen anything yet. The first thing was a major leap in technology: the introduction of the National Center for Supercomputing Applications (NCSA) Mosaic web browser in 1993 (and the subsequent founding of Netscape in 1994 by Mosaic team leader Marc Andreessen and entrepreneur Jim Clark). This innovation gave Internet users easy access to multimedia sources of information and popularized the

World Wide Web. The second thing was the award to Network Solutions of a cooperative agreement that would forever change the company.

NETWORK SOLUTIONS

Contact:
Mary Bloch
Mgr, Corp. Communications
(703) 742-4740

NETWORK SOLUTIONS AWARDED REGISTRATION SERVICES FOR NSFNET AND THE NREN

Herndon, Virginia, January 8, 1993 -- Network Solutions Incorporated, a leading network integration firm headquartered here, today announced it has received a cooperative agreement to provide registration services management for the NSFNet and the National Research and Education Network (NREN).

Network Solutions anticipates the value of the registration services award to be approximately $5 million over five years. Under the cooperative agreement, the company will serve as the worldwide registrar for the non-military portions of the Internet. Work will be performed at corporate headquarters in Herndon with a dedicated staff of eight to nine. Network Solutions has provided these services to the Internet community under a subcontract for network information services to the Defense Data Network for the past two years. The company will continue to support the Department of Defense community for the duration of the DDN subcontract.

In the spring of 1992, in cooperation with the Internet community, the National Science Foundation developed and released Project Solicitation NSF92-24 for one or more Network Information Services (NIS) Managers to provide and/or coordinate Registration Services, Directory and Database Services, and Information Services for the NSFnet. As a result of this solicitation, three separate organizations were competitively selected to receive cooperative agreements totalling over $12 million. Along with the Network Solutions award for Registration Services, AT&T was awarded the Directory and Database Services component of the project, and General Atomics of San Diego won the Information Services component of the project which is now named "the InterNIC".

-- more --

505 Huntmar Park Drive • Herndon, Virginia 22070 • (703) 742-0400

Network Solutions is a $35 million technology services company recognized nationwide as a multifaceted integrator of computer and telecommunications networks. The company specializes in internetworking, interoperability and network management. Network Solutions has a distinguished twelve-year history of excellent performance in the application of networking technology to support the missions of customers in government and industry.

-- end --

Figure 5-1. NSI news release announcing award of National Science Foundation cooperative agreement, January 9, 1993.

5

GAME CHANGER: THE NSF COOPERATIVE AGREEMENT

The Awardee shall provide to non-military internet users and networks all necessary registration services [which were] previously provided by the Defense Information Systems Agency Network Information Center (the DISA NIC).

— NSF Cooperative Agreement No. NCR-9218742,

January 1, 1993[87]

Because of its extensive work on the Internet and its working relationship with the National Science Foundation, in 1992 Network Solutions decided to bid on an NSF cooperative agreement to serve as the Network Information Services Manager for NSFNET and NREN and to provide InterNIC registration services. NSI group lead Scott Williamson and senior engineer Mark Kosters put together a proposal to provide these services to NSF—a role it had already been fulfilling for two years as a subcontractor to Government Systems, Inc., on its contract with the Defense Information System Agency.[88] NSI won the cost-plus-fixed-fee cooperative agreement (NCR-9218742) with a start date of January 1, 1993, and an expiration date of September 30, 1998.[89]

According to Mark Kosters, the working relationships the company—and in particular, Scott Williamson—had developed with NSF played a critical role in the InterNIC win. Says Kosters, "Scott had a good, trusting relationship with NSF, especially Don Mitchell. I'm sure that relationship was critical in our win of the registration services bid for the InterNIC."[90]

In the text of the cooperative agreement, NSF described the process it went through in making the award to Network Solutions (and to two other awardees):

> In cooperation with the Internet community, the National Science Foundation developed and released, in the spring of 1992, Project Solicitation NSF92-24 for one or more Network Information Services Managers (NIS Manager[s]) to provide and/or coordinate (i) Registration Services, (ii) Directory and Database Services, and (iii) Information Services for the NSFNET. As a result of this solicitation, three separate organizations were selected to receive cooperative agreements in the areas of (i) Registration Services, (ii) Directory and Database Services, and (iii) Information Services. Together, these three awards constitute the NIS Manager(s) Project.[91]

The estimated total amount to be paid to NSI under the agreement was $4,219,339. At the same time, two other cooperative agreements for Network Information Service Managers were awarded by NSF. One was awarded to AT&T (NCR-9218179) to manage database and directory services[92] (according to George Strawn, called the Yellow Pages and White Pages for the Internet within the National Science Foundation[93]), and the other was awarded to General Atomics (NCR-9218749) to provide information services.[94]

Specifically, the NSF cooperative agreement required NSI to perform four tasks related to providing Internet registration services:

- Domain name registration
- Domain name server registration
- Network number assignment
- Autonomous system number assignment[95]

According to George Strawn, NSF knew that Network Solutions was uniquely capable at the time to perform these Internet registration services, despite the company's small size. Says Strawn, "We knew they were a small company. The other thing we did know, though, is that, up to that time, they had been doing domain name registration for part of the DARPA activities, so they were experienced. DARPA gave them a good rating for their performance, and anybody else that we would have taken on for the job at that point would have been inexperienced with this activity."[96]

But why didn't the National Science Foundation keep these functions for itself—why did the organization contract out the work in the first place? According to Strawn, "NSF doesn't keep anything in house. NSF supports science and development, and it does not do science and development. NSF has no laboratories, like NIH or Energy or NASA. We provide assistance support for the research community. We made three cooperative agreements with the three awardees. We always look at what we do as not government business and, therefore, we don't do government contracts. We provide what we call assistance awards. And that was the case in all three of these examples."[97]

Don Telage was NSI's first president after the acquisition by SAIC, and he served as NSI's chief operating officer from 1995 through 1997. Before taking a position at Network Solutions, Telage was a group senior vice president and head of SAIC's networking group, and before that a systems engineering manager at GTE Strategic Systems Division where his focus was communications systems design and development.[98] In an interview with CNET News, Telage explained the impetus behind NSF's decision to push these functions to Network Solutions:

The National Science Foundation explained to me two of the dilemmas that were facing them at the time. One of them was that all the registrations that were coming in—or almost the majority of them—were commercial registrations. So there they were sitting with a federally taxpayer-funded contract to support commercial registrations, and they really felt very uncomfortable in that circumstance. The Internet had evolved underneath them. Furthermore, the usage numbers were growing, and the kind of capital that we would need to grow the equipment base and add to service offerings was way beyond the $5 million cap that that contract had for five years. So there was no way that they could figure out how to keep up with this growth and allow NSI to do the investments necessary to continue to meet the service requirements. A second problem they faced is that they had their first domain name trademark suit and they were terrified of being embroiled in litigation.[99]

Interestingly enough, Network Solutions almost didn't bid on the NSF cooperative agreement to serve as the Network Information Services Manager for NSFNET and NREN and to provide InterNIC registration services. Scott Williamson was the project manager for DDN NIC in 1991–1992 when NSI was doing domain registration services under a subcontract to Government Systems, Inc., in Chantilly. But when NSF put out the bid for the NREN project in 1992, which would separate the military from the nonmilitary names, Scott had to fight hard to get NSI's owners to bid on the NREN project.

For Network Solutions, which had somewhere between four hundred and five hundred employees at the time, winning the eleven-person NSF cooperative agreement wasn't really a big deal. In fact, it didn't take long for the agreement to start causing pain for the company's management team—perhaps making them wish they had turned down Scott Williamson when he introduced the idea of bidding on it.

The agreement was cost-plus-fixed-fee—a type of contract that reimburses a company's costs for project performance while guaranteeing an agreed-upon profit—and the total value of the agreement was capped at approximately $4.2 million over its five-and-a-half-year lifetime, or only about $764,000 a year.[100] Gary Desler describes the problems the company faced with its NSF agreement: "There might have been two hundred dot-coms on the Internet when the agreement started, and then along comes the World Wide Web, and we were suddenly getting two hundred requests for domain names a minute. NSF wouldn't let us staff up."[101]

At the same time, NSI was facing a new problem. As people began to realize that the web was going to be the next big thing, they started registering as many domain names as they could—whether or not they represented the company that actually had rights to the name by way of their registered trademarks. There were no rules in place about who could register a domain name, nor were there rules restricting the purchase of specific domain names. Concerned that people would buy up large blocks of domain names simply to resell them for a profit, Network Solutions attempted to institute a series of administrative roadblocks. Domain name purchasers would be allowed to have only one domain name per organization on .com. This restriction didn't last long, however. NSI also tried to restrict .org to nonprofits, and .net to organizations that dealt with Net infrastructure. Once the registry was auto-mated, these restrictions also went away.[102]

It didn't take long for trademark holders to realize that in many cases their domain names had already been registered by other parties. As a result, Net-work Solutions became the target for lawsuits—and threats of lawsuits—filed by companies and individuals that felt their domain names had been unfairly given out to others. After being sued by registered trademark owner Knowl-edgeNet, Inc., to recover the knowledgenet.com domain from a Virginia con-sultant, NSI finally put together a domain dispute process. This process gave preference to the owners of federally registered trademarks in disputes over domain names, while requiring that applicants indemnify Network Solutions for third-party claims that NSI was required to defend.

Still, the lawsuits continued.

When NSI went to the National Science Foundation for relief from these lawsuits, both actual and threatened, it was told that it was NSI's problem, not the government's. At the same time, the company was experiencing increas-ingly serious financial problems. The 1990–92 recession hit many companies hard, and Network Solutions was no exception. The company was growing fast, and it needed cash to support this growth.

There was a glimmer of hope when NSI was approached by a group of venture capitalists to see if there was any potential interest on the part of NSI's executive team to acquire a Sunrise (Ft. Lauderdale), Florida, technol-ogy company by the name of Internet Systems Corporation. The company was working on non-IBM interfaces to the Defense Data Network. The deal was this: the venture capitalists proposed that if NSI acquired Internet Systems Corporation, then they would put about $6 million into the company. So Gary Desler, Ty Grigsby, and a tech employee flew down to Florida to check out Internet Systems Corporation. They looked at the work the company had done, went through the offices, and interviewed the employees. The acquisi-tion proceeded, and Internet Systems Corporation was brought into NSI as a division.[103]

The venture capitalists were now in the possession of Network Solutions stock, but the promised $6 million cash infusion into NSI never happened. Six months later, after spending at least $500,000 on the venture, NSI was forced to trash the deal. Although the company was able to sell some of the products it developed (DEC/VMS & Concurrent OS/32 TCP/IP stacks), this outcome left NSI with a serious cash flow problem, making securing additional funding for the company's ongoing operations more critical than ever before.[104]

NSI's leadership team made the rounds of the nation's financial centers—New York, Philadelphia, Atlanta, and more—pitching the idea that the Internet was on the way, and NSI had some great ideas for leveraging its intimate knowledge of the Internet and track record of success for profit. According to Gary Desler, the response of these financiers was universally negative, "They'd say, 'What's the Internet?' and laugh."[105]

The problem was that, in the early 1990s, relatively few businesses were yet aware of the Internet, and even fewer were aware of the huge business potential for this global communications medium. Businesses had not yet caught up with this fast-moving technology, or with its possibilities. And while this situation would change in a very big way in just a few years, that did Network Solutions no good at this critical time in its growth and development. The company was unfortunately early to the party, and as such, it had little choice but to sit and wait for the rest of the world to arrive.

PART THREE

Too Small To Scale

W e live in a world of numbers and statistics, used for a variety of different purposes, including measuring the relative degrees of success—or the lack thereof—of businesses in this country. For example, the US Small Business Administration (SBA) has long kept statistics on the failure rate of small businesses. According to the SBA, seven out of ten new employer firms survive at least two years, half at least five years, a third at least ten years, and a quarter stay in business fifteen years or more.[106]

There are of course many factors that go into the complex equation of whether a business succeeds or fails. The size of the initial capitalization, the business acumen of the founding team, the ability of the company to get a viable product or service to market—and to gain significant sales—before the company's cash is depleted, the ability of key players to secure bank financing or additional investment along the way, the response of competitors to the entry of the company into their markets.

As you will learn as you read the chapters in this part, Network Solutions was in a challenging position. Not only was the company perennially undercapitalized, it rarely made a profit. And while it appears that NSI was able to provide its customers with good service despite these challenges, it's clear that the ownership team was under no small amount of stress.

Many observers report that undercapitalization is the number one reason for business failure. That may be true, but I do not believe that this observation goes far enough. In a recent article in the *New York Times*,[107] Chicago entrepreneur Jay Goltz reported ten reasons behind the failure of most small businesses. I personally believe that two of Goltz's ten reasons had particularly adverse impacts on NSI's fortunes:

- Out-of-control growth. This one might be the saddest of all reasons for failure—a successful business that is ruined by over-expansion. This would include moving into markets that are not as profitable, experiencing growing pains that damage the business, or borrowing too much money in an attempt to keep growth at a particular rate. Sometimes less is more.

- In the case of Network Solutions, the business expanded quickly over a relatively short period of time—even before the company signed the cooperative agreement with NSF that gave it the keys to the Internet domain name registry. And after that agreement was signed and domain name registrations began to flood the company, growth got out of control in a very big way.
- Lack of a cash cushion. If we have learned anything from this recession (I know it's "over," but my customers don't seem to have gotten the memo), it's that business is cyclical and that bad things can and will happen over time—the loss of an important customer or critical employee, the arrival of a new competitor, the filing of a lawsuit. These things can all stress the finances of a company. If that company is already out of cash (and borrowing potential), it may not be able to recover.
- Network Solutions did not have the cash cushion it needed to help it survive the combination of fast growth, operational inefficiencies, and slow customer payments. Instead, the company seemed to be on a constant search for new or increased bank lines of credit and investment from individuals and other businesses. The lack of a cash cushion became problematic as the company's debt grew, leading up to the sale of the company to SAIC by NSI's ownership team.

In this part, we'll examine the circumstances that led to NSI's ongoing financial challenges and why the company's ownership team ultimately had little choice but to sell the company. The lessons from this case are applicable to any business that is growing fast and that finds itself under stress as a result.

6

A FINANCIAL TAILSPIN

We were growing, but financially we were falling apart...We were doing bids for information services, network address assignment for cellular networks using IP, and that sort of thing. Meanwhile, the company was hardly making payroll.
— Mark Kosters, former Network Solutions senior engineer[108]

Unfortunately for Network Solutions and its principals, things were quickly going from bad to worse. In an effort to diversify its offerings, NSI decided to get into the computer hardware distribution business. With government agencies moving in a big way to networks of personal computers, the thought was that NSI could further leverage its position as a provider of computer networking services to the government by adding computer hardware to its product offerings.

There's an old saying that timing is everything. While I don't personally believe that the ultimate success or failure of a venture can be fully attributed to timing alone, I do believe that the timing was wrong in this case for Network Solutions to embark on this new product venture. Computer hardware at that time had become a commodity, and margins were becoming increasingly thin. So unless you had some extraordinary value to add to the hardware, there was little real money to be made by someone who wasn't the manufacturer. After spending about $1.5 million going down that particular blind alley, NSI pulled the plug on computer hardware distribution.[109]

In addition, NSI had signed two fixed-price contracts, one in Pennsylvania as a sub to Lockheed (the company has since been renamed Lockheed Martin) for a statewide computer network, and one in DeKalb County, a suburb of Atlanta, to network the school system.

In the case of the Lockheed contract, NSI's subcontract stipulated that the company wouldn't get paid until the prime got paid. Unfortunately, the prime—Lockheed—was a year behind on its delivery. Network Solutions had purchased large quantities of networking hardware and dipped into its cash reserves to pay for the project labor. The company was forced to sit on these costs for more than a year, while it waited for Lockheed to deliver and get paid by the state of Pennsylvania.

The DeKalb County contract required NSI to install local area networks in 112 schools. The county required NSI to subcontract out most of the actual installation labor on the contract to local firms, while NSI focused mostly on the engineering and electronic side. Unfortunately, the "typical" school that NSI was given by DeKalb County to use as a model for its bid turned out not to be typical. Most of the 112 school buildings had double firewalls in between classrooms, something the model school did not. The result was that the installation contractors refused to honor their bids to NSI. DeKalb County was stuck because it didn't have any extra money to put into the project that school year. DeKalb asked NSI to proceed, with a promise that the county would make good on payment the next school year. While the county did eventually pay, the damage was done.[110]

Network Solutions was struggling financially, and the pressure was building internally to do something—anything—to relieve the strain. Emmit McHenry had been out beating the bushes all along, looking for sources of financing, but he and his team could not secure a source of capital large enough or steady enough to put the company on a firm financial footing. NSI had received numerous queries from venture capitalists, but nothing ever seemed to pan out from these initial probes.

Regardless, Network Solutions was becoming more attractive to potential investors. The company had built a strong résumé in its computer networking capabilities, and its portfolio of government and commercial contracts was growing.

Emmit's long-term working relationship with Allstate paid off when the company's venture capital group made a small equity-based loan to Network Solutions. Kodak also made a small investment in the company, brought in by an investment-banking firm in Hartford, Connecticut, with which Ty had a relationship.[111] Kodak was for a time interested in acquiring NSI, because of the company's growing portfolio of government and commercial contracts. Unfortunately, because NSI was still in the SBA's 8(a) program and the government would not allow NSI's set-aside contracts to convey along with the sale of the company, Kodak decided against making an offer for Network Solutions, and it put its acquisition plans on permanent hold.[112]

Combined, the Allstate and Kodak investments amounted to about $5 million. But these millions of dollars were still not enough to sustain the company for long—a significant portion was used to buy out the venture capitalists NSI had picked up as a part of its acquisition of its Florida-based Internet Systems division.[113] The company always seemed to be a day behind and a dollar short.

Even after Network Solutions won the NSF cooperative agreement to provide domain name registration services—what would turn out to be the tech deal of the century—adequate financing could not be secured for the company. According to Emmit McHenry, "I found it impossible to get folks focused on the opportunity. Network engineering and the Internet didn't play well in 1993 and 1994."[114] Continues Al White, former vice president of corporate marketing for Network Solutions, "I guess people at that particular time didn't understand the Internet. It wasn't getting the press, so they didn't think it was such a big thing."[115] Emmit made his pitch to Wall Street, but the answer was still a resounding no. Says McHenry, "I went to the middle Wall Street guys. I wasn't networked well enough to go where the real money was. Had I gone to the big guys on Wall Street, I might have stood a better chance."[116]

As a stopgap measure, NSI turned to Maryland National Bank, which had recently opened a branch in Tysons Corner, Virginia, to provide a loan against the company's accounts receivable. The bank would loan Network Solutions 75 or 80 percent against a government receivable and age it up to ninety days. On day ninety-one, NSI would take it out of the pool. While this loan helped keep the company in business, the hemorrhaging of cash was still going on—gradually building up to about $5 million of debt.[117]

Says Mark Kosters, "We were growing, but financially we were falling apart. At that point we were leaving the government 8(a) contract set-aside fold and not doing so well. We knew we were on the forefront of the Internet boom, and we were desperately trying to find other ways of gaining revenue within the company. We were doing bids for information services, network address assignment for cellular networks using IP, and that sort of thing. Meanwhile, the company was hardly making payroll."[118] NSI's banks were getting involved in the day-to-day operations of the company, telling its owners which computer equipment and other purchases NSI could make—and which it could not.

In the first year of the cooperative agreement, NSI registered thirteen thousand domain names and experienced a net loss of $386,000. In 1994, the financial loss for the year more than doubled—to $980,000.[119]

Mark Kosters was forced to run the entire InterNIC off of a single, two-gigabyte hard drive, which has only about 0.2 percent of the space of the one-terabyte hard drives widely available today. This despite the fact that worldwide use of the Internet was quickly mushrooming, and NSI badly needed more computer storage space on which to house all the registry information. Eventually Kosters was able to scrounge up an additional two-gigabyte drive, but it was a bare disc with no housing or any other hardware to install it into the computer. Says Kosters, "There was no additional money to do anything, so I just placed that disk loosely into the machine and hoped and prayed that it wouldn't crash. It never did, and it was the registry for the Internet for years."[120]

In 1994, SAIC made an offer to acquire Network Solutions that addressed the specific needs and desires of each of the company's four key owners. And while NSI's owners didn't at first jump at this offer, we kept the company on the front burner and kept the heat on high until we achieved our goal.

7

THE OWNERS SELL OUT

If you go back to what it was in the early 1990s, the Internet was not much more than a science fair project. It's a real testament to SAIC because there are very few companies on the face of the globe that could have taken this little mish-mash of servers sitting in college closets and hardened it into something that could handle the exponential volume and demand curve that NSI experienced.

— Bill Roper, former CEO, VeriSign[2]

When Network Solutions entered into a cooperative agreement in 1993 with the National Science Foundation (NSF) to develop and run the Internet's domain name registration service, it instantly became a government-blessed monopoly with the sole power to assign the .com, .org, .net, .edu, and .gov addresses that route users around the Internet. While few at the time foresaw the remarkable growth the Internet would go through just a few short years later—nor the tremendous amounts of revenue and profits that this growth would generate—Network Solutions was clearly in the right place at the right time. And once it became a part of SAIC, it gained the right team of people to take it to a level that few ever imagined would be possible.

Years before NSF's award of the cooperative agreement to Network Solutions in 1993, the company was already on the radar screen of an executive at Science Applications International Corporation (SAIC)—Mike Daniels. In the spring of 1987, John Woods, the lawyer at Patton Boggs who had been retained by NSI—and who, coincidentally, had done legal work for Mike—said, "I'd like you to talk to some fellows who are running a small company called Network Solutions. I'm their outside counsel, and I would like you to meet them because you've done some acquisitions and they're thinking about doing a first acquisition."[122] This first acquisition was for Sunrise, Florida-based Internet Systems Corporation.

Mike Daniels was involved as a user of the Internet from its beginning in 1969, when he was a young naval reservist on active duty, assigned to the Office of Naval Research in Washington, DC. According to Mike, "They assigned me to ARPA—it just so happened that the ARPANET was started a few doors down the hall from my office. So for the next two years, I was one of the first users of the ARPANET when it went into operational use."[123]

After he left the navy, Mike attended law school and, after graduation, took a job with CACI International—one of the early federal government information technology contractors—in Arlington, Virginia, before he left in 1979 to start with a partner a government-focused IT company: Computer Systems Management. His interests turned back to the Internet after Computer Systems Management was acquired by SAIC in 1986, and Mike joined SAIC's senior management team.

So, in early 1987, Mike met the four primary owners of Network Solutions: Emmit McHenry, Ty Grigsby, Ed Peters, and Gary Desler. Says Mike, "I immediately liked Emmit, Ty, Gary, and Ed, and I talked to them about the acquisition they wanted to make. I looked over the acquisition papers, gave them some advice, and they ended up buying a small software company in Florida. I kept in touch with them, and from 1987 to 1992 I would see Emmit or Ty once in a while and we would talk about things—they actually did a little business with SAIC, so I got to know them and the company."[124]

Mike Daniels continued to follow Network Solutions and keep in touch with its principals. He also kept a close eye on where the Internet technology was going—attending NSF conferences on the development of NSFNET. He was intrigued by the new Internet technologies funded by DARPA and NSF—such as the Mosaic web browser—that were emerging from places like the University of Illinois, Champaign-Urbana, and finding their way into the Silicon Valley. Mike knew that Network Solutions had done a little work with NSF and DARPA and, while talking with Emmit McHenry, he learned that NSI was interested in bidding on a cooperative agreement that the NSF was going to release for Internet registration services.[125]

Meanwhile, there were ongoing discussions within SAIC's management team about how we were going to grow from a billion-dollar company into a $2 billion, $5 billion, and someday $10 billion company. We thought about getting into the oil and gas business because of the company's established relationships with BP, but we also considered getting into the commercial transportation business, the telecommunications business, and the Internet business. I assigned Mike the task of looking into the Internet business, and he brought forward the idea of acquiring Network Solutions.

Says Mike Daniels, "Bob Beyster had asked me to look into the Internet business, which I took as looking into how we could potentially get SAIC into it. So it was a very serendipitous confluence of events—the relationship with Network Solutions that I had continued since 1987, going to a few NSF meetings, talking with some people from the Silicon Valley. While most of Emmit's business was with the federal government, he'd won a few networking contracts with large organizations like Nations Bank and the state of Pennsylvania. I told Bob about Network Solutions, and I put together a briefing and took it out to California, and Bob and I talked about it."[126]

I was impressed with Mike's briefing, mostly for Network Solutions' well-established computer and telecommunications networking business, but also for its involvement in the Internet. I was aware of the Internet because of my association with Steve Lukasik and others on the SAG—the Strategic Advisory Group for the Department of Defense. Lukasik was continually demonstrating to us the wonders of what was then the ARPANET, and the military was tremendously interested in it. So the more I talked to Mike about Network Solutions, the more I thought it would be a worthwhile acquisition candidate. We started getting serious about the prospects of a deal with NSI's owners, and I told Mike to talk to Emmit McHenry about buying the company. Mike and his team started the process.

Bob Korzeniewski, at the time SAIC's corporate vice president for mergers and acquisitions and eventual CFO for Network Solutions from 1996 through 2000, recalls the courtship of NSI's four owners: "So we went and visited with Emmit McHenry and his three other partners—I think the first time we visited with them was June 1994. It was a really different situation—each of them had a different perspective on the company. Emmit clearly ran the company and made all the decisions, but Gary Desler, Ty Grigsby, and Ed Peters were also part of this four-person ownership team. So we had to figure out what was important to each of them. By the time we eventually closed the deal, each one of those guys had a different deal, which is really unusual."[127]

SAIC made a series of offers to NSI's ownership team, but for one reason or another, a deal remained elusive. According to Mike, "During 1994 and early 1995 we made Emmit four or five offers for the company. And each time Emmit would say he was interested, but maybe the price wasn't right, or he wasn't ready to sell, or he was worried about how his people would be treated if they became SAIC employees. So we went back and forth."[128]

As time went on, the financial pressure within Network Solutions to get a deal done steadily increased. The company continued to burn cash, its banks continued to hold a tight leash on spending, and NSI was losing what little leverage it had. In addition, NSI was becoming involved in lawsuits (such as the knowledgenet.com case mentioned earlier) as a direct result of its domain name registration business. This brought additional pressure to bear on NSI's owners. Fighting these legal challenges required considerable political and legal muscle—something that NSI just didn't have at the time. And by then, the company couldn't afford to buy the muscle it needed to get the job done.

Finally, Mike Daniels and I were able to break the logjam. Recalls Mike, "I told Emmit that I wanted him and Bob Beyster and me to have dinner. We had dinner with Emmit at Morton's Steakhouse in Tyson's Corner, and Bob talked to him about why he should come and join SAIC. Emmit said he would think about it and let us know in a day or two. He called me the next afternoon and said, 'The only reason I'm going to sell you the company is because I've gotten to know you and I trust you and I trust Bob Beyster and whatever you guys tell me, you'll do—otherwise, I would never sell you the company.'"[129]

On March 10, 1995, SAIC bought Network Solutions, Inc., for $4.7 million in SAIC stock. SAIC also assumed about $5 million of NSI's debt. According to Bob Korzeniewski, "Network Solutions had two big investors that had lent them money—they had a huge amount of debt and credit that we were able to get down. We paid them about twenty or thirty cents on the dollar to get the deal done. What ended up being a pretty inexpensive deal for SAIC in real-dollar terms had underneath it the complexities of a much larger deal. On top of that, I think it's always hard for someone to sell his company. It was especially hard for Emmit, but he was a class guy throughout the entire process."[130]

After the buyout, the four previous owners of Network Solutions went their separate ways. Emmit McHenry went on to found a new business, NetCom Solutions International, Inc.—a telecommunications, engineering, consulting, and technical services company—where he served as chairman and CEO.[131] Original company founder Gary Desler joined SAIC, where he remained employed in a variety of different technical positions until June 2010.[132] Ed Peters also initially joined SAIC, but he eventually left to start his own company. Tyrone Grigsby helped found Nehemiah Project International Ministries (NPIM) of Clackamas, Oregon, a business-training organization that provides biblical-based entrepreneurship training and support for small- to medium-sized businesses around the world. There he eventually became board chairman, a position he retains to this day, although he still lives in the Washington, DC, area.[133]

Although paying $4.7 million for a company that had rarely made a profit since it incorporated in 1979 seemed like an awful lot of money, I saw great promise in Network Solutions' networking business and the potential for NSI to get us more deeply involved in the commercial network market. This is ultimately what convinced the SAIC board to approve the acquisition. However, Mike Daniels felt that the Internet business might eventually grow into a worthwhile venture, and he turned out to be right. Regardless, we both felt strongly that NSI represented a good business opportunity for SAIC at the time, and we decided that acquiring Network Solutions was the right decision.

Don Telage remembers when he first became fully aware of the opportunity that Network Solutions and its small contract with the National Science Foundation presented us. Says Telage, "I'll never forget that Mark Kosters was the guy who woke me up to the potential of that little million-dollar-a-year contract with the National Science Foundation. He kept hounding me to come down to his basement office and to show me some growth curves. I was busy with a billion other things, but he finally dragged me down there and showed me an exponential growth curve—it was very clear to me that something amazing was happening with the Internet."[134]

What was happening with the Internet was that personal computers were getting a significant foothold in the marketplace—in businesses and in people's homes. So personal computing had become a reality, and sales of personal computers were growing fast. At the same time, a decent web browser in the form of Netscape had been introduced to the public in late 1994. This product allowed people who weren't computer experts to easily access the World Wide Web. In addition, wired connectivity—in the form of telephone modems—began to become widely available, enabling computer users to gather information and communicate with others—next door, or around the world—through a variety of online services, which quickly grew in popularity.

Finally, the growing availability of online services such as America Online, Prodigy, CompuServe, and others provided users with a variety of software applications and communications and information resources, including messaging, e-mail, message boards, chat, games, news, weather, and much more, and quickly popularized the online world. As these early services transformed into Internet service providers—and as freestanding websites began to proliferate—the true potential of the web for individuals and businesses began to be realized, and growth rocketed upward.

All of these different events converged at about the same time—in about 1995—which happened to be the year we acquired Network Solutions, Inc.

After the acquisition, we absorbed NSI's government business into SAIC and left the company's commercial business where it was. We then began an internal debate about what to do next with this little company. We knew we had something interesting in the Internet business, but NSI's small cost-plus-fixed-fee contract with the National Science Foundation was no moneymaker. As often happens in life, while we considered our options, events quickly caught up with us—propelling both SAIC and NSI into entirely different and much higher orbits.

PART FOUR

MONETIZING THE WEB

U nless you are running a nonprofit organization, at the end of the day it's all about making money for your company's owners, shareholders, and investors. They put their money at risk in the hope that you will be able to provide them with a greater return than the many other investment options available to them at any given point in time. If you can do that, then more investment dollars will follow. If you cannot, then those dollars will move elsewhere—to someone who can meet or even outperform projections.

When we acquired Network Solutions in 1995, the Internet was for the most part still a noncommercial entity. The government, specifically the National Science Foundation, was still subsidizing the cost of domain name registrations, and there was little money to be made anywhere in the system. The first ISPs had emerged about five years earlier, and they were eking out a growing flow of revenues and profits selling commercial Internet access to companies.

This all changed when we at SAIC joined with NSI's leadership team to convince the government to allow Network Solutions to charge a fee for domain name registrations. The idea was approved by NSF—indeed, the agency was anxious to put the money it was spending to subsidize domain name registrations to other uses—and NSI started charging $100 for a two-year domain name registration. With the sharp increase in domain name registrations during the mid-to-late 1990s, the fee (lowered in 1998 to $70 for two years) resulted in huge infusion of cash into NSI's coffers and a business that was extremely profitable. While NSI cannot claim to be the first business to commercialize the Internet, its revenue model pointed the way for the many Internet businesses that followed, including the slew of dot-coms that started operations in the late 1990s. However, unlike most of those dot-coms, NSI actually made significant profits after the acquisition by SAIC.

The federal government and the private sector have a long history of partnering to move technology and innovation forward in this country. NSF, for example, has long been an important supporter of technological innovation, with the agency playing a key role in the development of numerous technologies. Some of these technologies include bionic retinal implants, deep-sea research vessel *Alvin*, bar codes, computer-aided design and computer-aided manufacturing (CAD-CAM), optical fiber, magnetic resonance imaging (MRI), and much more.[135]

Despite the pressures of tightened government budgets, there is still great opportunity for private-sector companies to monetize technologies developed or funded by the government. In fact, as budgets continue to tighten in coming years, I believe there will be increasing pressure for agencies to quickly transition their technologies to the private sector, not only to minimize ongoing outlays for further development, but to spur our nation's continued economic development from products that may result.

In this part, we'll consider how Network Solutions was able to monetize its part of the Internet, while fighting off challenges from entrenched, old-line Internet users who thought the Internet should not be commercialized (and who often fought against commercialization), and politicians who wanted to exert power over the increasingly lucrative and important American Internet industry.

8

PAY TO PLAY

According to industry observers, the $50 fee is unlikely to cause much controversy, because it's relatively small. "This is like charging nickels for a Coke in the company vending machine. [It's] a token amount," said Jerry Michalski, managing editor at Release 1.0, a newsletter based in New York.

— Article, *InfoWorld*, September 18, 1995[136]

Our top priority for NSI was to turn the company around financially and put its bottom line firmly in the black. This meant figuring out how to monetize Internet domain name registrations.

Mike Daniels knew Steve Case and Jim Kimsey and had watched them build the predecessor to America Online, and he and the SAIC management team had seen Netscape go public. So, at the end of 1995, Mike went to Silicon Valley and spent the better part of a week talking to a number of lawyers who had been involved in the Netscape deal and other potential Internet IPOs.

A great deal was learned, but then everything changed.

In late 1995, the NSF granted Network Solutions the authority to charge customers a fee to register their domain names. Up until that point, the NSF had subsidized the domain registration service—a practice that inevitably had to change as the number of commercial registrations began to grow and the costs to support them began to climb to a level that would soon be unsustainable given NSF's budget constraints. As George Strawn recalls, "We had observed requests for domain names going through the roof. The cooperative agreement with NSI called for increased payment to the company if there was increased business to do, which was only reasonable. We made some back-of-the-envelope estimates and realized that, whereas we had budgeted about a million dollars a year for the cooperative agreement with NSI, if the increases kept coming at the rate we projected them, it would cost us more like a million dollars a month instead of a million dollars a year. We didn't have $12 million that we wanted to invest."[137]

In a January 1995 meeting, a couple of months before NSI's acquisition by SAIC, the NSF put Network Solutions on notice that the company would need to start getting cost recovery of its domain name registrations and that charging a registration fee might be the best way to go. Don Mitchell told Scott Williamson and Mark Kosters that the company needed to start charging for domain names and that it was going to happen soon. Scott and Mark put their heads together to determine the price that should be charged for each domain name, eventually coming up with a figure of $50. Jon Postel, who at the time was in charge of the Internet Assigned Numbers Authority, agreed that $50 was a reasonable number, and the plan moved forward.[138] According to Network Solutions founder Gary Desler, the idea to start charging for domain names had been in the works for two years before NSF finally granted this authority.[139]

On September 13, 1995, NSF issued an amendment to its cooperative agreement with Network Solutions—now a wholly owned subsidiary of SAIC—to establish an annual fee of $50 per domain name registered, or $100 for two years.[140] NSI thought that the fee wouldn't be a hurdle for most users—by then there were thousands of people and companies coming to the company each week to register their domain names. As it turned out, NSI's guess was right. Aside from the very vocal complaint within the traditional Internet community about having to pay a fee where previously domain name registration had been free, the company didn't encounter price resistance for the fee that was decided on. I believe this was because the value proposition at the time was so high. The Internet was quickly becoming a real moneymaker for the nascent online businesses, and $50 a year was a small blip in their larger financial picture.

Establishing this new domain name business required the ultimate skill in negotiations with many stakeholders, but the federal government—seeing the potential of this new information technology—helped push this financing strategy forward. Although NSF knew the Internet community would not be happy with this new fee arrangement—people were used to not paying for anything on the Internet at that time—based on the huge increase in domain name registrations beginning to hit the system, the agency was running out of options. The NSF cooperative agreement with NSI called for increased payments to NSI if there was increased business to do—money the agency didn't have. And the commercial interests that were beginning to flood domain name registrations were outside of NSF's scope, which was to support the registration of universities and scientific researchers.

NSF was essentially left with just two choices—taking care of this issue the Internet way, or taking care of this issue the expedient way. Going the Internet way at that time would have meant first telling the Internet community what the problem was, discussing it thoroughly for a year or more, and then making a decision. As this process meandered forward, NSI would potentially be burning through a million dollars a month. Going the expedient way would short-circuit the deliberative process of the Internet way—running the risk of angering many in the Internet community—but it could be executed immediately, saving the agency millions of dollars.

Ultimately, NSF made the choice to save millions of dollars.

According to George Strawn, who approved the amendment to the cooperative agreement with NSI, "That would have cost us $12 million. I didn't want that coming out of my budget, and I don't think anybody at NSF wanted it coming out of my budget, either. So I said, all right, I'm going to approve this, and we're going to foist it on the Internet community, and they'll be shocked and they'll be mad. A, they'll be mad that we didn't tell them. B, they'll be mad because we created an NSI monopoly. And C, they'll be mad because we put a surcharge on it. We estimated that the cost would be $35 a year, and we instructed NSI to charge $50 a year rather than $35, with the extra $15 going into what we called the 'Internet Intellectual Infrastructure Fund.' It was actually a two-year contract, so it was $30 out of $100 for two years."[141]

Says Don Telage about NSF's situation, and its approval of the domain name registration fee, "There were a lot of commercial entities getting on the Internet, so putting more federal funds into supporting the domain name registration function didn't make sense. Also, they had no money for infrastructure improvement. They knew this would be a growing problem as the

Internet took off. So the National Science Foundation was desperately looking for a commercial organization that had the stature and the financial backing needed. Network Solutions alone wouldn't be enough—that was clear. But Network Solutions backed by SAIC met NSF's needs."[142]

In the meantime, domain names and the registry controlled by NSI were quickly growing in importance as businesses increasingly realized that they needed to establish a beachhead on the Internet. NSF originally anticipated that it might get a couple hundred thousand dollars out of the Internet Intellectual Infrastructure Fund, which would allow the agency to privatize the last few dimensions of Internet support that NSF and a few other government agencies were doing. It would enable NSF to get out of the Internet business and move on to its many other responsibilities. The fund quickly grew to almost $30 million, a factor of one hundred times greater than NSF originally anticipated. This made it a political hot potato until Congress came to NSF's rescue and passed a law that instructed the agency to use the proceeds to fund the next-generation Internet project, which was underway at that time.[143]

Indeed, the addition of a fee to register domain names was a hugely unpopular move within the Internet community. Says Vint Cerf, "That was actually quite an important change, and I recall a great deal of intense debate on this whole subject of charging. NSF I thought had a perfectly understandable argument—it's spending valuable research dollars on something increasingly used by the commercial sector, and the commercial sector should pay for it. Nobody knew quite what the right prices were, so $50 was the nominal figure. I assume that the idea was to have a nominal figure until they could figure out what the real number should be."[144]

And many Internet users—accustomed to getting a free ride on the information superhighway—were quite vocal in their opposition to the new domain name fees. According to Don Telage, "We had to announce the registration policy change quickly because there was a leak—we hadn't yet gone public and someone found out about it. So we announced it at the last minute, and it was pretty ugly for a long time. I had unbelievable e-mail traffic sent to me, threatening e-mail traffic. I was terrified to go to my car late at night. We were crossing wires with a culture that thought the Internet was their playground. It was a very closed club, and it was very anti-commercialization, and

From Commerce...

...to E-Commerce

Figure 8-1. Cartoon: From Commerce to E-commerce, 1999. Reproduced with permission from John Cole.

we were in the middle of that at that point doing something that was heresy to them."[145]

The purist Internet community liked the way things were done, and they were not happy with the push to commercialize the Internet. In my opinion, this was a naïve perspective about how the world works. It wasn't 1970 anymore, and the sandbox they once owned was now a national—indeed, a global—asset. Although the community of Internet users didn't appear to object to the portion of the domain name fee that was being set aside for the Internet Intellectual Infrastructure Fund, it did object to the lion's share, which was going directly into NSI's for-profit coffers. They also objected to the fact that the National Science Foundation had granted NSI monopoly status in its Internet registration services duties. These objections would continue to build over the coming months, until they eventually spilled over—leading to fundamental changes in the scope of Network Solutions' duties.

9

REBUILDING THE DNS INFRASTRUCTURE

It's often said that no one owns the Internet. But a little company tucked into a red-brick office park here comes pretty close. For the past five years, Network Solutions, Inc. has held an exclusive contract to handle distribution of addresses for the World Wide Web, including its most widely recognized feature: the ".com" suffix. Currently appended to more than two million Web addresses, ".com" covers the pre-eminent domain for doing cyber-business. If Web addresses were California real estate, Network Solutions would own the coastline.

— *The Wall Street Journal*, 1998[146]

When SAIC acquired Network Solutions from its four-person ownership team, NSI had three main business areas: federal government contracting, commercial consulting, and domain name registration. Soon after we bought NSI, we took steps to reorganize it. We absorbed the federal government contracting business into SAIC—leaving NSI's domain name registration business and its small commercial consulting unit where they were. The company was managed independently of and separately from SAIC, and we initially sent over about five employees (including Mike Daniels, who took over as NSI's CEO and chairman, hired his own management team, and set up an independent board of directors) to join NSI's approximately 380 employees.

In addition, we initially provided administrative services to Network Solutions. These services were provided under arms-length service contracts between SAIC and NSI.[147] We also allowed the company to buy the hardware it needed to strengthen the InterNIC infrastructure.

According to Mark Kosters, "One of the big issues we had before NSI was sold to SAIC was competing for dollars. I was very outspoken, as was my boss Scott Williamson, that NSI was making a mistake funding initiatives that did not seem to make sense and providing minimal funding to us. Creating and staffing a Microsoft helpdesk was one such mistake."[148] By the time we came on the scene, there was much work to be done.

One area that desperately needed upgrading, for example, was the database that Network Solutions inherited from SRI International when the company picked up the NSF cooperative agreement. SRI had used this database since the late 1980s for DDN billing, and the version we had to work with had had very few improvements from what SRI was using. In early 1991, NSI decided to move from the proprietary SRI database to Ingres during the transition from SRI to NSI. However, the original database administrator was not versed in relational technology, and the initial database design was extremely poor. This caused performance issues, as it was hard to migrate to a more proper relational model given that NSI had a large code base that used the substandard schema.[149] As a result, it was difficult to search for customer records and to get information quickly.

The Internet prior to 1995 was, for the most part, the domain of technical people, some academics, and the military. All of the information with regard to domain registrations was mainly used for technical purposes—it had not been used much for commercial purposes or by individuals. This meant there wasn't really a need to keep close track of customer records or to do database searches quickly—it was all pretty casual. That situation changed dramatically in 1995 through 1996 and going forward. All of a sudden, Network Solutions had a real domain name business to run, generating real cash, with domain expiration dates and renewals to track. Customer records were important, and the ability to retrieve them and keep them up to date and organized was of great interest—making the development of a new and accurate customer database a high priority.

Soon after the announcement was made to NSI's employees that SAIC had acquired the company, Don Telage told Mark Kosters that money was no longer an issue for NSI. Says Kosters, "I explained to Don what we did and how we were going to make money. He said, 'Mark, I no longer want you to worry about money. I want you to buy whatever equipment you want and we'll make it happen.' So we went ahead and did that, with a pretty massive cash infusion from SAIC. We started building the InterNIC, and we started going down the path of setting up the service so we could start charging for domain names."[150]

Regardless, even the crown jewel of Network Solutions' assets—the A server—remained in a vulnerable position. Gabe Battista officially became NSI's CEO on November 1, 1996,[151] and he served in that position through December 1998, when he left NSI to take the helm of long-distance telephone company Tel-Save.[152] Before he accepted the chief executive officer position at Network Solutions, Battista served as chief executive officer, president, and chief operating officer of Cable & Wireless, Inc., in Vienna, Virginia. And before that, he held management positions at US Sprint, GTE Telenet, and the General Electric Company.[153]

Soon after he was brought on as CEO, Battista took a tour of the company's Herndon facility. Says Battista, "It was a small company at the time. Network Solutions was already choking on its growth, and I think one of the reasons that SAIC offered me the job was they felt that my operational and marketing experience could help put processes in place, consistent with the vision that someday Network Solutions could go public and be a major player in the Internet. But I didn't realize just how much there was to be done."[154]

Continues Gabe, "When I first got there, I think it was Chuck Gomes who took me downstairs to see the A server. It turned out that the A server was on the first floor of a glass-enclosed commercial building, with hardly any lock on the door and wires everywhere. Chuck was explaining to me, 'This is the A server, and there are eight other servers...' And I asked him, 'Well, what happens if there's a fire? What happens if a guy drives a car through the glass window? What happens, what happens, what happens?' And that was only because in the telecom business, when you build a switch somewhere, you make sure that you have fire protection, theft protection, and you bury it under the ground. Nobody knows where the switches are and it's safe and secure and nobody can get into the room. You've got to use ten different cards to get access to the room that has the switch. The first thing I realized is, oh my God, we've got to protect this server, which we did. We were lucky enough to bring on Dave Holtzman. After interviewing a list of candidates, I hired him. Dave joined the company, and he and his team were able to make the A server secure."[155]

Dave Holtzman's background was quite unique. Holtzman was a former cryptographic analyst and submariner with the US Navy, and he also worked at DEFSMAC (Defense Special Missile and Astronautics Center) as an intelligence analyst, focusing primarily on the Soviet Manned Space program.[156] Holtzman next served as chief scientist at IBM's Internet Information Technology group, where he managed the development of IBM's information product and service offering to encrypt and sell digitized content across the Internet. He served as a senior analyst for Booz Allen Hamilton for several years, where he ran technology-driven restructuring initiatives for Wall Street firms and large financial institutions, and he designed and built a networked, heterogeneous database and text retrieval system called Minerva, which was used by NATO and several trade associations before being sold to IBM in 1994.[157]

We on the SAIC board took our responsibility to secure the A server—and thus the Internet—quite seriously. If Network Solutions needed to improve the hardware installations—to make sure the system worked twenty-four hours a day, seven days a week, without fail—then we needed to give them the money required to do so. We authorized NSI to buy the latest computers, mainframes, servers, and PCs. We eventually spent tens of millions of dollars duplicating the A server installation, which was the Internet's true home. We committed SAIC resources to stand behind Network Solutions, and this commitment ensured its ultimate success.

Despite the infusion of cash from SAIC, it took Network Solutions some time to recover from its long period of undercapitalization. As the volume of domain name registrations ramped up starting in 1995, NSI's limited staff worked long hours to try to keep up. In just one example, Mark Kosters routinely pulled eighteen-hour days, six days a week, for eight years—from 1991 through 1999. Staffing was intentionally kept lean in hopes of generating more profit from the company's contracts. Some contract job slots were left unfilled and existing staff worked many extra hours to cover them.[158]

Says Kosters, "We had numerous issues. We struggled with the load— there were only two of us there essentially, me and one other guy. For a long period of time I was not only doing development, but also operations and responding to RFPs. I would spend my day doing engineering and public-facing things for the organization, then come back late at night for another round of engineering. And then, when I was almost too tired to think, that's when I would actually sit down and do registrations until four in the morning."[159]

And even with some of the most talented DNS experts on the NSI team, mistakes were made. For example, there were times when NSI's technical team would inadvertently take someone off the system, and once they introduced a bug into the Ingres database by accident—almost losing the entire domain name registry. (It took some quick work by Mark Kosters to recover all the data.) And then there was the time in June 1996 when Network Solutions shut down MSNBC.com, an event that became another worldwide newspaper headline. According to Mark Kosters, MSNBC.com was marked as paid in the Microsoft Access database. But the data was not marked as paid in the main Ingres database because of a Microsoft Access bug.[160] This led to the shutdown.

In an e-mailed apology to MSNBC, Network Solutions stated, "Please accept our apology for deactivating your domain name. Your payment was received prior to the cutoff date, but because of a technical error still being researched, the payment information was not properly applied."[161] At the time, NSI's billing system was still mostly manual and therefore labor intensive—causing a large backlog.

A significant reason behind NSI's billing challenges was the fact that many domain registration records were not up to date. Either the registrant information was out of date, the admin contact was out of date, or there was no billing contact—or sometimes a combination of the three. Network Solutions started charging for domain names on September 13, 1995, but the company didn't start mailing out invoices until four or five months later. And when it mailed out the invoices, many of them were returned because the addresses were out of date. In June of 1996, NSI's management made the decision to start putting domain name registrations on hold—that is, they wouldn't be active in the zone files. As soon as NSI's technical team inactivated a domain in the zone files, any websites or e-mail attached to the domain would no longer work. Putting a domain on hold definitely got peoples' attention, and in some cases it was the only way NSI could get in contact with a domain name registrant because its contact information was out of date. This is what led to the situation with MSNBC, as well as many other examples of less significance.

Aside from these issues, the chronic understaffing at NSI had created a problem with the domain name registration process. The backlog to register a domain name had climbed to six weeks, creating a significant lag between the time someone called to register a domain name and the time he or she could get a new website online.[162] And then there was the additional challenge of creating an entirely new and potentially complex billing system and back-office organization for the $50 annual domain name registration fees that had been approved by NSF.

But billing wasn't the only part of the domain name registration operation that was experiencing significant challenges. Network Solutions had a hard time just keeping track of the enormous influx of applications for new domain name registrations, much less billing for them. Mark Kosters knew that this problem was just going to get worse as the number of applications continued to grow, so he created a new system to deal with it. Says Kosters, "So I wrote this thing called a message tracking system. What it did was it actually queued up the applications until someone could get to them—that way they wouldn't be lost. And, actually, I think this is one of my proudest achievements at Network Solutions. The system scaled like nobody's business. When I first wrote the system, we were seeing numbers on a monthly basis. Later, this changed to a per-minute basis. All we had to do to handle the increase was just add more machines to work on the queues, and it did the job. If we hadn't had that system, we would have failed."[163]

When NSF issued an amendment to its cooperative agreement with Network Solutions on September 13, 1995, to establish the annual domain name registration fee, NSF expected that NSI would start collecting it right away. There was only one problem with this expectation: there was no billing system in place to collect the domain name registration fees and no mechanism to transfer $15 of each $50 registration fee into the Internet Intellectual Infrastructure Fund and then track and disburse it.

As Bob Korzeniewski recalls, "I think Don Mitchell at NSF called on a Thursday and said, 'You have to start billing on Monday.' We had zero back office—we couldn't send out an invoice. To be able to send out invoices, we had to do the accounting side of it, which meant we had to estimate how many of these people were going to pay. I don't think we sent out our first invoice until January or February 1996. So we had two or three months of domain names coming in, but no billing going out. That put us in a big catch-up mode."[164]

Those first few years after the acquisition were particularly challenging because we had to first work with, and then entirely replace, NSI's legacy customer service systems as the company tried to service an ever-growing workload. It was a bit like trying to rebuild an entire aircraft—while you were at 45,000 feet, flying from Los Angeles to New York.

NSI had boxes and boxes of returned e-mails and invoices that didn't get to the customers, because up until the company started charging for domain names, it wasn't nearly as critical to keep the domain name contacts up to date. As a result, NSI ended up sending out invoices to e-mail addresses that no longer existed because the people hadn't updated their records. Literally thousands of invoices didn't reach the registrant or their admin contact.

Network Solutions had a system that was, from an operational viewpoint, ill-suited for the job that the company was being asked to do. NSI had just one customer when it took over the contract from the National Science Foundation. NSI sent the customer a bill and NSF paid it. It went from that to literally millions of customers within a few short years.

In those days, there were no systems to do automated customer call management like there are now—which integrate data files and phone calls so that you can hand them off to people. There was no online automated fulfillment system. Network Solutions didn't have an automated online billing system. NSI did have some of the largest online customers in the world, and they struggled to manage the registration service. Another challenge was that the intentionally non-proprietary TCP/IP system didn't interface with the proprietary IBM Systems Network Architecture that all the mainframe companies like First Data were using for their back office. NSI was engineering the customer-maintenance, customer-care, and customer-billing systems in order to do business with the prevailing large-volume fulfillment houses. No interfaces between the Internet and these systems existed at the time.[165] These had to be fixed.

And if that wasn't enough, NSI was getting phone calls from all around the world—the domain name business was a global one—and the company's staff wasn't prepared to deal with customers who didn't speak English. It wasn't huge in terms of percentages of customers, because the primary growth at that time was still in the English-speaking world, and the Internet at that time was made up primarily of English speakers. But NSI's management team did start to hire more people with different language skills, and that became helpful—both for NSI and for the company's foreign customers. Network Solutions didn't address the language issue very well for many years, and this became more critical as time went on.

Gabe Battista understood the technical challenges that NSI's customer service people had to deal with because of its legacy systems, which were archaic and not user-friendly. It was difficult for the company's customer service reps to quickly and easily find the information they needed to help customers. When a customer asked the customer service rep a question, he or she would have to dig to find the information. All the registrations were e-mail based at that time—there was no web-based system. In fact, there was no computer-based customer service system at all. To answer customer questions, NSI's customer service reps had to sort through the sole technical database and e-mail archives. It was a time-consuming, cumbersome task.

Given the circumstances, the Network Solutions team did a commendable job, but it didn't look good to the outside world. People wondered why NSI couldn't be more responsive—especially given that customers were now paying $50 a year for a domain name. NSI couldn't exactly go out and tell them that the company was trying to do the best it could with its antiquated administrative systems. But that was the reality of the problem. And it was many years before those legacy systems were replaced. In the meantime, Network Solutions hired lots of people to deal with the systems it had until the systems were eventually upgraded.

The National Science Foundation's insistence that NSI take checks in payment for domain names created another major and ongoing headache. That was something NSI's management team didn't want to do. The team wanted to make credit cards the only payment option, but it was one of the hard negotiating points in the meetings NSI had with NSF before July of 1995. The National Science Foundation insisted that NSI have the ability to take checks. This caused the company untold grief.[166]

Recalls Don Telage, "Probably the most frightening experience that I had at Network Solutions was one day I went down to the room where we were handling the payment and billing systems. I opened the door and took a look in the room where people were working. There were stacks of boxes of what looked like computer punch card boxes—the long ones that are about three inches tall—up against the wall, from floor to ceiling about twenty feet long. I said to Chuck, 'What the hell are those boxes?' He said to me, 'Checks we don't know how to apply.' There were boxes of checks—tens of thousands, maybe even hundreds of thousands of checks that didn't have the domain name that we could apply them to. But we had to apply each one to some-body's account to do a billing reconciliation. We eventually applied fifty peo-ple to the problem and it took twelve months—making phone calls, writing e-mails, doing whatever we could—to figure out which domain name each check should be applied to. One by one, box by box. It was a nightmare."[167]

In November 1996, NSI was reincorporated in the state of Delaware,[168] and on March 11, 1997, Network Solutions announced that it had registered its one-millionth domain name. At the same time, news articles loudly pointed out that NSI had collected fees on potentially only about half of these domain name registrations. According to one report, "Network Solutions is required to bank 30 percent of the fees it collects in a trust fund, but as of the end of January, that fund had only accrued $15 million, or 30 percent of $50 million—half of what a million registrations would net."[169]

Indeed, rather than banking the $100 million in domain name registration fees it was entitled to if every payment for the one million registrations had been collected, the company had apparently only collected about $50 million.

As vice president, Internet Relations and Compliance for Network Solu-tions, Chuck Gomes served as the primary interface for the NSI Registry to the Internet community and also oversaw contract compliance. Gomes had been a part of Network Solutions' management team since 1984, previously serving as vice president, Customer Programs, and he managed NSI programs and projects involving delivery of technical services to a variety of state and federal government agencies.[170] He also played a key role in developing a lasting solution to the company's growing backlog of registrations and invoices.

Understandably, creating a stable and efficient billing system for the domain name fees—with up-to-date contact information—became a top priority for Network Solutions. Mark Kosters ended up developing a prototype of the system at home on his laptop computer, using the Microsoft Access database software. It was during one of these homework sessions that Mark began to realize that Network Solutions was onto something very big with the domain registry. "I wrote the billing system in Access because I wanted to be home with my wife, and laptops at that point didn't support Linux or any sort of UNIX distribution. I remember sitting down with her while I was working on the prototype billing system—she never really understood what I did. In fact, she was very much annoyed that I was never home. However, on the TV there was a commercial for something.com, and I said to her, 'Oh my goodness, did you see that?' She said, 'What are you talking about?' I said, 'There's a domain name—they registered that through us!' I then realized that this thing was turning a corner."[171]

To get some idea of how this growth in business had an impact on the NSI organization, consider the following example. In 1995, one month after NSI started charging for domain name registrations, the company ran the entire InterNIC project with just twenty-five people (including co–project managers Mark Kosters and Chuck Gomes, and Kim Hubbard, who headed up the IP allocation team at NSI for nine years in her position as IP registration manager). Within the next couple of years, staffing for domain name registrations exploded, with three hundred employees devoted just to customer service.

It took time to turn the corner. While we at SAIC invested more money in upgrading NSI's systems than did its previous owners, we were cautious and conservative in our spending—as we were in all our business operations. It took time for us to see operational and financial returns because we were hiring people, making investment in equipment, and securing new facilities. We did eventually help NSI secure the equipment it needed to allow customer service people to answer the phones efficiently, pull up the information needed at work stations, and monitor customer service to make sure NSI's customers got the right information.

Dave Holtzman and his team did a great job hiring the right operational people, obtaining new equipment for NSI, and upgrading facilities. The efforts of Holtzman and his team were a key turning point for Network Solutions, and for the ability of the web to scale up as user demand mushroomed. Without the attention they paid to upgrading the system's hardware and facilities, growth of the network could have been seriously constrained.

So Network Solutions did eventually get caught up, and even with the company's explosive growth, NSI was able to handle its business. The Internet was clearly the wave of the future. Businesses had found it, but they didn't understand how essential the basic infrastructure of domain names was to the Internet's operation and just how precarious it was. Volunteers ran many of the root servers around the world, and these servers were not well secured. NSI at one time did a survey of name servers. Most of them were in public places. The I server in Sweden—the first root name server to be established outside the United States—was in its early years operated by students at the Royal Institute of Technology in Stockholm.[172] The D server sat in the middle of the University of Maryland student library.[173]

Recalls David Holtzman, "We were inundated with crises pretty much two or three times a week. I can remember the data center flooding a few different times because of broken air-conditioner lines. We were the first organization to put any security at all onto any critical piece of the Internet other than the actual cables themselves. Part of the reason we did this was in response to all the crazy phone calls we were getting. We got death threats. We got bomb threats. We got people calling almost every day screaming, mostly because they were angry because they wanted a domain name and we wouldn't give it to them—typically because someone else had gotten it first. It was not uncommon to see a domain selling in the high six figures or even for a million dollars or more. And when you're talking that kind of money, people take it pretty darn seriously."[174]

The tragedy of 9/11 was a wakeup call to this nation and its security in many different areas—including the protection of name servers and other key parts of the Internet infrastructure. There are a lot of people today who don't want the Internet to go away—who can't afford for it to go away—and who depend on it to do business. We knew in the late-1990s that if the Internet was going to be a viable long-term platform for business, it had to be made more secure, it had to be managed, and it had to be a process. So Network Solutions continued to invest in and improve its domain name systems and administrative and technical processes, and we at SAIC continued to support it.

One of the most valuable things that we in SAIC did as a company was educate the powers that be in Washington. Most of the people in Congress, the White House, the military, and the federal agencies did not understand what the Internet was, nor did they understand the increasing importance of the domain name registration system and the threat to our nation if the system for any reason was disabled or collapsed for a prolonged period of time. They certainly didn't understand the technology, but they slowly began to understand the importance of the network to American—and even global—business. We spent countless hours educating government decision makers on why it was important for them to pay attention to Network Solutions and its registration business. And I believe we were for the most part successful in our efforts.

Ultimately, we were successful in our efforts because once we educated business and government decision makers who were starting to use the Internet, they saw that the Internet had to in some way be managed, be secure, and have a process. Although the Internet came out of a casual and loose university environment, the tide turned in the direction of formalizing systems and controls and making it more suitable for doing business. We saw the value in Network Solutions; we bought it; we did our best to revitalize the company; and we built a great relationship with the National Science Foundation. The NSF liked that we brought the A server and other DNS infrastructure under firm control. What the NSF didn't like was the ongoing and often contentious struggle we had with the people and organizations arrayed against us.

People began to realize that the Internet was a gold mine, and the rush was on. At the same time, the Internet's creators worried that the network would be taken over by businesses and commercialized, and that it would fundamentally and forever be changed from the free-and-open model that they had originally envisioned. And Network Solutions was right in the absolute center of it. Not only did it own the keys to the mine, but it also operated the sole tollbooth on the increasingly busy highway in and out of that mine. At the time, if you wanted to establish your .com presence on the Internet, Network Solutions was the only game in town. And we were a large lightning rod for the many people who did not want the Internet to be commercialized. But now, there was no turning back from commercialization.

According to Gabe Battista, "I think one of the things that ultimately resolved the issue was, as all contentious issues are resolved, is you have to find a win-win. And one of several key parties that had to win was the US government, because it felt like it wanted to protect this future platform for business. The international community had a stake in it because it was going to be a worldwide web. They had feelings about it and wanted to win. The business community had a vested interest, and they wanted to win—they wanted to be secure. They didn't care who secured the Internet, but they knew that somebody should secure it. And then of course we wanted to win, as did the Internet Society and the founders of the Internet such as Jon Postel."[175]

As of the first quarter of 1996, Network Solutions had registered approximately 246,000 domain names. However, by the second quarter of 1999, the cumulative number of domain name registrations initiated by NSI had mushroomed to 5.3 million.[176] One of the results of this swell in domain name registrations was an enormous surge in revenues and income for the company, which was keeping $35 of the $50 annual registration fee for itself and sending $15 to the Internet Intellectual Infrastructure Fund. This influx of cash enabled NSI to make key hardware purchases and plans to move into a larger facility that would allow the company to grow the systems required to support the dramatic growth of the DNS. In July 1997, Network Solutions closed a deal on an additional 31,247 square feet of space located in Herndon's Sugarland Business Complex at 365 Herndon Parkway. The term of the lease was negotiated to run from May 30, 1997, through July 31, 2002, and the stage was set for NSI's continued growth.[177]

Net Revenue

(milllions of dollars)

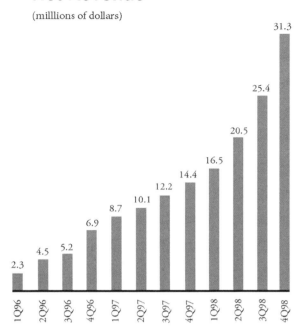

Figure 9-1. NSI net revenue growth, 1Q96 through 4Q98. *Source*: Based on Network Solutions, Inc. Annual Report (1998). *Graphic by:* Jamie Dickerson, The Foundation for Enterprise Development.

While I don't personally believe that SAIC's purchase of Network Solutions in 1995—as it was spiraling downward due to its severe financial problems—saved the Internet, it certainly went a long way toward building the foundation required for its future growth and worldwide success. Absent our acquisition of NSI, someone would have likely stepped up to take over the domain name registry—perhaps SRI International would have taken back the responsibility from the National Science Foundation, or maybe the National Science Foundation would have found another company qualified to run the system. But there is no guarantee that the results would have been as positive as they were under SAIC.

Net Income
(thousands of dollars)

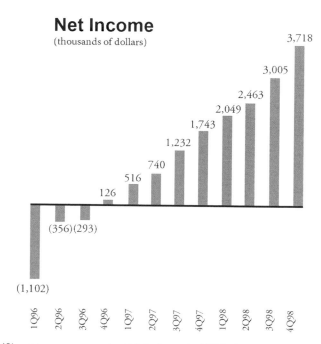

Figure 9-2. NSI net income growth, 1Q96 through 4Q98. *Source:* Based on Network Solutions, Inc. Annual Report (1998). *Graphic by:* Jamie Dickerson, The Foundation for Enterprise Development.

George Strawn told me in an interview, "If you or somebody else hadn't bought NSI and put as much resources into it as you did, given the unanticipated rate of growth, the entire thing would have melted down. So I've always viewed SAIC as one of the unsung heroes of the Internet, keeping it afloat and working. I rested a little easier knowing it was getting top-level attention in the organization."[178]

Ty Grigsby also believes that SAIC had a major and positive impact on NSI's effectiveness as an organization and its ultimate success. According to Ty, "I think what SAIC did as a company after they bought Network Solutions from us was nothing short of amazing. They understood the political and legal ramifications of the projects they were involved in. I don't think a lesser company could have pulled this off."[179]

Not only did our involvement in the Internet during the course of the short five years we owned Network Solutions do a lot toward stabilizing the DNS and creating the infrastructure necessary for its future growth, we helped to create a model for companies to charge for providing Net-based products and services. In the final analysis, this latter contribution may very well have been the most important one. According to David Holtzman, "I don't believe that Network Solutions was a technology powerhouse. In fact, I've often said to people that as important as my job was, it was probably one of the least

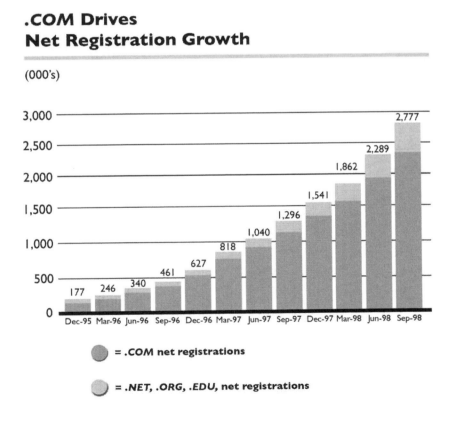

Figure 9-3. .com Drives Net Registration Growth. *Source*: Based on February 1999 Network Solutions, Inc. Prospectus. *Graphic by:* Jamie Dickerson, The Foundation for Enterprise Development.

technical jobs I've ever had in my career. But what was important for both myself personally and for Network Solutions was policy and management, and that's where we made the difference. And in 1995, there was no policy. No one was able to charge for domain names, and by extension there was no financial infrastructure across any of the Internet other than vague access charges, which were even then fairly ill-defined. Network Solutions was able to lay down a process for a business model for the Internet."[180]

So Network Solutions did begin to build up the infrastructure and to invest in capital equipment to support the growth of the domain name system and the Internet as a whole. NSI hired more people to deal with the manual systems it had to deal with in supporting customers, and the company got more involved in Internet policy activities and participating in meetings. But while the infrastructure that Network Solutions put in place to support the InterNIC and DNS was vastly improved within the first couple of years after the acquisition, not everything was perfect.

David Holtzman remembers one occasion when the entire domain name system almost came to a grinding halt. Says Holtzman, "There's this piece of software called BIND (Berkeley Internet Name Domain, developed as a University of California, Berkeley, graduate student project), which is the resolver software that the DNS system uses. So that when you type in 'www.ibm.com,' at some point it goes into this program called BIND. It runs on most big computers, and it translates 'www.ibm.com' into a TCP/IP number. Well, BIND had always had a lot of problems, and one particular problem in this case was that it had a limit on the length or the width of a number that it could handle. If I remember correctly, it was an unsigned integer or something like that. So there was some hard-and-fast limit, which we were about three days away from hitting. If we had hit that limit in the software, the entire DNS would have just stopped. So we went out and spent some money and upgraded to a 64-bit instead of a 32-bit computer and quickly ported the code—surviving to another day. That's not engineering—that's using chewing gum to fix a hole. But we did that very well."[181]

But while this particular landmine was thwarted at almost the very last minute, NSI didn't always lead a charmed existence. This was proved in a very public—and, for some, ominous—way one warm summer night in July 1997.

10

THE NIGHT THE INTERNET DIED

Ignored Warning Leads to Chaos on the Internet.[182]
> — *The New York Times*, July 18, 1997

Human Errors Block E-Mail, Websites in Internet Failure; Garbled Address Files From Va. Firm Blamed.[183]
> — *The Washington Post*, July 18, 1997

Internet Glitch Reveals System's Pervasiveness, Vulnerability.[184]
> — *The New York Times*, July 18, 1997

During the 1990s, when the Internet was rapidly growing and increasing in importance in the day-to-day functioning of businesses, governments, and other organizations, I don't believe anyone ever imagined that the entire system could go or would fail on a widespread basis. After all, the Internet was by design a distributed communications system, with more than 220 root name servers strategically situated around the world. If one root name server—or even two or three—went down, then the remaining servers would automatically step in, causing just a momentary blip for Internet users as the traffic was quickly routed around the disabled servers.

At least it was supposed to work that way. One day, the unimaginable could and did happen. And it happened on our watch.

During the late evening of July 17, 1997, at about midnight, a technician at Network Solutions loaded a file onto the company's A server—the computer that contained all the millions of domain names on the Internet. This routine task was performed to update the Internet with the new domain names added each day, and there was no reason to expect that this update would be any different from the thousands of daily updates that had preceded it—all without incident. Unfortunately, on this particular day, the NSI technician loaded the A server with an empty, null file instead of one with all the world's domain names. As other servers around the world propagated the file, the Internet began to shut down, region by region, until most of the system was dead.

For the majority of its relatively short life, the Internet had been little more than an electronic mail system for moving messages quickly within government and academic technology circles. But by 1997—with the invention of the World Wide Web (or simply the "web" for short) in 1990, and the release of the Mosaic graphical browser in 1993—businesses had increasingly jumped onto the Internet bandwagon by buying domain names, building websites, engaging in e-commerce, and using the network to send critical business data across the nation and around the globe. So the complete crash of the Internet on that night in July 1997 didn't pass unnoticed.

Within minutes, websites could not be accessed, e-mail couldn't be sent or received, and critical data flows were frozen. According to one article, by the time NSI got the system working properly again, "countless thousands or even millions of e-mail messages had been returned as undeliverable, while untold numbers of users had been unable to make contact with various World Wide Web sites whose addresses were temporarily garbled."[185]

Mark Kosters recalls that fateful night. "We were generating this thing called a zone file—that's where all the names for the database were actually thrown into DNS so that people could resolve their names. It crashed when it was creating the .com zone files. So there was no data in there whatsoever for anything that ended in .com, .net, or .org. We had a safety mechanism in place to say, 'Hey look, something is wrong here. Don't go forward until you get someone to approve it.' But at that point there were so many domains no longer being paid that still needed to be removed from the system, that the operators were getting fairly careless."[186]

What had happened was that the safety mechanism to prevent the installation of a null file on the A server had been routinely ignored by the operators, in much the same way that many people today ignore a blaring car alarm. While overriding the automated safety mechanisms had never before led to any sort of significant problems—based on their past experience, the operators didn't have to do anything about the tagged files—in this case history was not an accurate predictor of the future. The operator installed the null zone file, overrode the failsafe switch, and immediately pushed the file out to the thirteen other root servers around the world, which duly accepted it without question as they always had. Almost instantly, nothing was resolved in .com, .net, or .org.

Figure 10-1. Photo of the AMS-IX mirror of the K root name server. Reproduced with permission from Bas van Schaik.

At the same moment that the Internet was melting down as a result of the NSI technician's error, the system was dealt another serious blow. Just hours after the NSI error, a backhoe operator working near Laurel, Maryland, severed a major East Coast fiber-optic cable, disabling five hundred high-speed data and voice lines owned and operated by AT&T, Wiltel, Sprint, and MCI. As repairs were made, Internet traffic was rerouted, and the net effect was a significant slowdown in the system. The combination of the fiber-optic cable loss along with the corruption of the domain name system caused Internet traffic to grind to a virtual halt for at least a day in much of the United States, and even longer throughout the rest of the world.

Continues Kosters, "At about four in the morning, I got a call at home from a user in Germany, he said 'The Internet's down, it's not working—what's going on?' My home number was listed in the phone book, and I figured people knew how to get a hold of us—we had lots of kooks call us over time. My wife picked up the phone, and she was not very happy because it was yet another night where I had a service call in the middle of the night that woke us both up. So I went downstairs and checked—the Internet was in fact down and we needed to fix it. I got to the office and had it fixed by six in the morning, but by then the damage was done. Starting at about nine in the morning, the press coverage became intense. We had people calling us from all over the world asking 'What happened?,' 'What did you do wrong?,' 'How did this happen?,' and so on. It made front-page headlines of most the major newspapers around the world."[187]

David Holtzman served as chief technology officer of Network Solutions and the manager of the Internet's master root server. He adds, "One of my system admins was on honeymoon in Hanoi, and he brought back a Vietnamese newspaper. The only part I understood was 'Network Solutions,' and I could guess what the rest of the article was. I started getting heavily involved with the media."[188]

Gabe Battista had become CEO of Network Solutions in October 1996, and he was at the company's helm when the Internet went down. On the evening of July 17, 1997, Gabe was preparing for a vacation to Italy. Gabe's well-laid vacation plans were soon overcome by unforeseen events. Recalls Battista, "At about 2:00 a.m., I got a call at my home from Dave Holtzman. He said, 'The Internet's gone down.' I asked, 'What do you mean the Internet's gone down?' Of course, Dave meant exactly what he said—the Internet was down. Somebody loaded a null file—an empty file—into the A server, and the effect was cascading around the world and dropping all the domain names along the way. People couldn't get onto the Net. And that afternoon I was supposed to be leaving for Italy on vacation."[189]

Gabe called a good friend, Dan Dutko, who was at a lobbying firm—Dutko and Associates—in downtown Washington, DC. Gabe had known Dan for years, and Gabe knew that he was talented dealing with people on the Hill and in the administration. Continues Gabe, "I said, 'Dan, I have a big problem—the Internet's gone down,' and I told him everything. He asked, 'Where are you now?' and I told him I was in my house—it was 6:00 a.m. by then. Dan said, 'Don't go anywhere, I want to call John Podesta.' I had met John Podesta before—he was chief of staff for President Clinton. About an hour later, I got a call from John Podesta. He was very direct and straightforward. He asked me to tell him what happened, how long it was going to take to fix the problem, and what we were doing to make sure it wouldn't happen again. I told him, 'Here's what I know, and here's what I think.'"[190]

Gabe received one more phone call from John Podesta later in the day to see if things were progressing like he thought they would be—they were. So, as the day went on, Gabe still had to decide on whether he should go to Italy. In the meantime, he was having nonstop conversations with Bob Korzeniewski and Mike Daniels. As it became clear that the Network Solutions technical team had things under control, Gabe decided to go ahead and make his planned trip to Italy. He arrived in his hotel room in Rome at about eight in the morning, and turned on the television. Says Gabe, "The first thing I did was put on CNN. And the headline on CNN is 'Internet goes down—caused by a small company in Herndon, Virginia, called Network Solutions—it did something wrong.' I said to myself, 'Oh my God.'"[191]

The death of the Internet brought immediate and negative publicity to NSI, and by association, SAIC. Almost every website, after all, was shut down to many users around the world. According to the *Washington Post,* David Graves, Internet business manager for Network Solutions, said of the mistake, "What occurred…wasn't a technical problem. It was a human problem. We don't know what was going through the individual's mind—and we're taking appropriate action."[192]

The root problem was fixed within an hour, and the Internet was gradually brought back on line in a matter of days, with the entire system functioning normally once again within a week. Imagine trying to explain to a horde of agitated members of Congress, White House staffers, and the world's press corps why the Internet was broken and how long it would be before it was fixed. These were questions that Network Solutions honestly didn't have the answer to at the time. And it was a graphic demonstration of just how dependent people, businesses, government, and academic organizations of all kinds were becoming on this fast-growing electronic network. It showed the criticality of what the company was doing and the necessity to do it like landing an airplane, not right 99 percent of the time—or 99.5 percent of the time—but right 100 percent of the time.

Recalls Dave Holtzman, "The thing that this incident brought home to me was how quickly everything started to melt down. Even though we fixed the problem in about half an hour—forty-two minutes, if I remember correctly—the repercussions lasted for days and days and days. It was amazing to me that a small thing like that could have such an enormous impact, and that it would happen so fast and affect so many people. From that moment on, I've been frightened because the basic architecture and the infrastructure of the Internet hasn't really changed at all since 1997. They still push the zone files out. There's no security per se, no electronic security involved in it. There's a lot more money thrown at it, but the basic systems are still the same and what happened then could happen now except of course, the scale would be orders of magnitude greater. Imagine, hypothetically, if the stock market crashed on a Thursday or Friday and it started dropping through the floor. Now also imagine that the Internet flamed out in the middle of this market crash. The result could be financial—and economic—chaos for our nation."[193]

11

TURNING THE CORNER

Network Solutions (NSOL) this morning jumped into the public arena with a strong showing, as its first trade hit the market at nearly 40 percent above its target price. The domain name registrar, which launched its initial public offering at a time when Congress is debating its future, entered the market with a first trade of $25 a share—up from the $18 a share price underwriters set last night.

— CNET News, September 26, 1997[194]

As it turned out, NSI was at the nexus of several converging technologies and social circumstances. The World Wide Web had been launched—first in Europe in 1990, then in the United States in 1991. The development of the Mosaic/Netscape browser soon followed—in 1993—with a user interface that allowed regular people to get onto the Internet. These events were coupled with a large increase in the penetration of personal computers, both in businesses and in the homes of much of the public. Internet service providers began to spring up, providing anyone who had a phone line and a modem to plug into the Net.

And—as the gatekeeper to the Internet, for good and bad—Network Solutions and its domain name business were at the epicenter of it all.

After the acquisition, we at SAIC began the process of taking a closer look at what exactly it was that we had and what we might do with it. It started to occur to us in 1995 that the commercialization of the Internet was in full swing and that there was the potential for significant shareholder value to be created. We had seen Netscape go public on August 9, 1995, and watched as its stock soared from an initial offering price of $28 a share to $75 a share on the first day of trading.[195] Mike Daniels was personally acquainted with Steve

Case and watched as he and Jim Kimsey built Quantum Computer Services, Inc.—the forerunner to Internet giant America Online (AOL)—just a couple of blocks down the street from SAIC's large office complex in Tysons Corner, Virginia. We knew that we now had possession of the NSF cooperative agreement by way of Network Solutions and that there might be some way to commercialize this piece of the Internet.

At the end of 1995, Mike went to Silicon Valley and spent about four days talking to lawyers who had been involved in the Netscape IPO and who were working on other Internet IPOs. He also talked to people in Silicon Valley companies that were trying to commercialize the different pieces of the Internet that they had. Mike was introduced to several different law firms that were working for these companies. One of the attorneys Mike met was Jorge del Calvo, a partner at the Pillsbury law firm in Palo Alto. Del Calvo's focus was on the representation of public and private technology companies in mergers and acquisitions, public offerings, private placements, and joint ventures. Jorge and Mike spent four or five hours talking about the latest Silicon Valley business trends, with the commercialization of the Internet dominating the conversation.

Says Mike about the results of his trip, "When I came back from that trip, I was convinced the commercialization of the Internet was underway and it was really happening in Silicon Valley. Then I spoke with Steve Case here in Washington, and we talked about what they were doing to get AOL off the ground in its early days. And I talked with George Strawn at the National Science Foundation—we spent probably four or five hours over several weeks talking about what the agency had been doing and what they would like to do in the future. So I learned a lot about what NSF's vision of the future was."[196]

So Mike and I started to talk about the possibility that the Internet industry would continue to grow at a high rate and that the domain name business we had in Network Solutions would start to take off. We realized that if these events came to pass, then we would have quite a different kind of animal on our hands than we had originally suspected. If the domain name business did start to take off—and in late 1995/early 1996, we didn't yet know whether or not it really would—then we would consider running Network Solutions as a commercial company that would focus its energies on the domain name business.

Mike and I started to talk about these opportunities, and we started to talk to people inside and outside of SAIC to gauge their opinions. We eventually came to the realization that this indeed might happen—the domain name business was on the verge of going exponential, and we would need to be ready for a massive surge of business. This would require bringing new people into Network Solutions, and it might require transforming it from a wholly owned subsidiary into a separate entity of some kind.

This brought us to the next step: considering exactly what this new commercial company would look like and how the internal systems would work. Although we had some commercial business, we were primarily a government contractor—an entirely different animal from a purely commercial business. How would the benefits work? What kind of accounting system would we have to put into place at NSI? How would the company track customer orders and payments? We realized that we couldn't just muddle along with a patchwork of SAIC systems, because NSI and SAIC were two completely different kinds of businesses. So I told Mike, "You know, you ought to start talking to people who might know how we might do something like this. You ought to start talking to people who might finance something like this."

Mike recalls, "So in 1996, I did two key things on that front. First, I started zeroing in on the financing question. I went to Silicon Valley and I asked Jorge del Calvo who he thought the very best bankers in Silicon Valley were for taking high-tech companies public. I flew to New York, where I met with the investment bankers. I met with J.P. Morgan. I met with Morgan Stanley. I met with Goldman Sachs. I met with Lehman Brothers and a variety of Silicon Valley boutique firms, including Montgomery and Company, Hambrecht and Quist, and others. Second, I knew we better start looking for some commercial people, because if this really happened, we would need them. And sure enough, in 1996 as the domain name registration business started to grow, our biggest problem was finding the right commercial people for Network Solutions."[197]

As a result of Mike's extensive meetings with technology company executives and owners, and their attorneys and investment bankers, the writing on the wall was becoming quite visible. Network Solutions was on the launch pad and prepared for liftoff. We knew that we would succeed only if we prepared ourselves for a huge wave of business that was rapidly growing, with no end in sight. This would require that we take Network Solutions public in an IPO so that we could raise significant funds needed for this anticipated growth.

Things were moving fast—both within Network Solutions and in the Internet community as a whole. In July 1997, President Clinton issued a presidential directive on electronic commerce, making the Department of Commerce the agency responsible for managing the US government's role in the domain name system.[198] This effectively ended the National Science Foundation's involvement with the administration of the domain name system and with Network Solutions, and it required NSI to develop and cultivate an entirely new set of relationships within the federal government.

Says George Strawn about the reasons behind the transition of Internet oversight within the federal government, "It was clear that NSF was not the proper agency to hold this contract—if the contract had to be held, it wasn't us. The Clinton administration did not want the FCC to hold the contract, since that would smack of regulation, and the administration was four square behind the deregulated Internet. So, the Department of Commerce became the appointed figure. We then entered into a year or two of discussions to get them to agree to pick up the responsibility for overseeing the Network Solutions activity and broadening the number of corporations that provided network services and so forth. We were in agreement with Ira Magaziner that this was a good thing to do."[199]

According to Mike Daniels, the change of government responsibility for the DNS didn't happen overnight. "The genesis of the transfer of responsibility from the National Science Foundation to the Department of Commerce was the ongoing discussions we had with NSF in the person of George Strawn and others. As the domain name business started to increase dramatically, NSF's position was, 'We're not sure we're the right people in the government to be involved in this because it's not our role to be part of a large commercial activity.' I had a number of meetings with a variety of people within the National Science Foundation and the Department of Commerce, with White House senior adviser on policy development Ira Magaziner, and twice in the West Wing with Vice President Al Gore. The White House decided that jurisdiction of domain names within the government should be transferred from NSF to the Department of Commerce in the person of Clinton's then Secretary of Commerce, Bill Daley. The right-hand man who we worked with literally for two years to accomplish this transition was Andy Pincus, who was general counsel at the Department of Commerce."[200]

If Emmit McHenry and the other owners of Network Solutions had not decided to accept SAIC's offer to acquire the company, I am convinced that the Internet would be very different from the way it is today. We brought in a large infusion of cash, along with some of the most talented technical, business, legal, and legislative minds available anywhere. Not only that, but we brought with us a name—Science Applications International Corporation— a trusted brand that was well regarded in government circles. The decision makers within the government believed NSI knew how to do the technology involved in running the DNS, and that if the company had gaps in its understanding, we would fill them with talented people.

And they were right. We at SAIC had the ability to bring in the financial and other resources needed to get the job done, while the pre-acquisition Network Solutions did not. Our mindset was, let's do the job right or let's not do it at all. In the case of Network Solutions, we were 100 percent committed to doing the job right. We didn't acquire NSI to patch it up and then flip it for a fast profit. We bought Network Solutions as part of a long-term strategy to strengthen our computer and telecommunications networking experience and employee roster. Indeed, we absorbed most of the non-Internet part of Network Solutions into SAIC. Soon after we acquired Network Solutions, there was a tidal wave of Internet adoption that drove the huge demand for names and, ultimately, websites.

Bill Roper served as SAIC's chief financial officer and a senior vice president from 1990 to 2000 and played a leadership role in Network Solutions from the time of its acquisition by SAIC in 1995 until its acquisition by VeriSign in 2000. According to Roper, "From a business standpoint, very few companies on the face of the globe probably could have taken this little mish-mash of servers sitting in college closets around the globe, kind of loosely hooked together, and hardened that into something that could handle the volumes as they grew, because this thing grew exponentially during the time that we were in charge of it. Shortly after we bought Network Solutions—in 1995, I think—there was a little party there that I just happened upon, a cake-and-cookies kind of party. It was the first week we'd ever registered one thousand names in a week, and everyone was celebrating the occasion. A year and a half later, we were registering something like two million names a quarter. That's a pretty steep curve. But the system worked, and it worked well. And it got hardened, both under the SAIC regime and then later under the VeriSign regime. So to me, that's a real business accomplishment. We didn't create the demand. We didn't really do anything but respond to it and not fail."[201]

David Holtzman also believed that SAIC was the right company at the right time—with the right people and technology—for this critical turning point in the evolution and growth of the Internet. According to Holtzman, "Out of all the other stakeholders, I don't know any other group that was involved at the time or involved today that I think was as well-motivated and as practical-minded as we were. I believe that one of the strengths was the fact that SAIC was involved. Generally speaking, we did the right thing—the right thing being we opted in almost every case in favor of keeping things stable and functioning. This was not rocket science. We weren't out to create brand-new, intellectually interesting things. We were out to keep things functioning. We were an engineering-driven organization because of the SAIC ownership. And I think in every case I had cooperation from the board, from Dr. Beyster on down. I certainly had to prove myself on a number of occasions, but if I could make my case strongly enough, I never had a problem. If we had been a marketing-driven company like many other companies, that wouldn't have happened, and I think the history of the Internet would have been a very different thing."[202]

That said, due to our lack of experience in commercial product businesses, we surely missed or mishandled a variety of opportunities to further expand NSI with product marketing and migration to other emerging areas of Internet products and services. The rapid revenue growth we experienced was entirely due to market demand, and we did little to stimulate it. We merely responded to that demand in the best way we could, combining the demonstrated technical expertise of NSI's staff with SAIC's deep pockets and political acumen.

Looking back, I don't believe Network Solutions would have been able to accommodate the exponential growth of domain name registrations without SAIC's direct help and involvement. And I don't believe the overall system would have been as stable as it turned out to be without SAIC's direct help and involvement. Without the ability to deal with the massive growth in domain name registrations, and without the maintenance of a stable and reliable system, the entire domain name system might have been severely hobbled, or even paralyzed, for perhaps months or even years. If this had come to pass, the National Science Foundation would have been forced to dig NSI and InterNIC out of the mess. Ultimately, the government may have had little choice but to take back InterNIC from the company and run it internally or send it back to SRI International. Had this occurred, I believe that the commercialization of the Internet would have been significantly slowed, putting the brakes on the dotcom boom and greatly attenuating the remarkable amounts of value created by Internet businesses during that period of time.

And I am certain that the cash-strapped company would not have been able to fight off the fast-mounting legal and legislative challenges that buffeted Network Solutions in the 1996 to 1998 timeframe. The real value of SAIC to NSI was our own technical capabilities and the financial credibility and capacity to ensure that the domain name system infrastructure could be stabilized and made robust in time to survive the coming torrent of user growth, which exceeded 20 percent a month for several years. I believe that SAIC is one of the few companies on the face of the earth that could successfully handle this challenge. We at SAIC had a deep bench of talent from which to draw and enough cash in the bank to pay for it. This is something that the pre-acquisition Network Solutions just did not have.

Network Solutions had other problems to deal with besides the "nice problem to have" of extremely rapid growth. The Internet community was lobbying the government hard against the new domain name fee and against NSI's monopoly position as both the registry and registrar of domain names for the Internet. In addition to these distractions, the company had to deal with a surge of lawsuits filed by companies and individuals that were convinced Network Solutions had wronged them in some way. This made the next couple of years very interesting indeed.

PART FIVE

STAYING ONE STEP AHEAD OF THE POLITICAL MACHINE

F or most of its formative years, the Internet was completely off the radar screens of Washington, DC, politicians and government insiders. As former VeriSign CEO Bill Roper described the scene at the time, "If you go back to what it was in the early 1990s, the Internet was not much more than a science fair project."[203] And because it wasn't much more than a science fair project, few people in positions of power within the government cared or had even heard about it.

That all changed when Network Solutions began charging a fee for domain name registrations, and the gold rush that was the Internet in the late-1990s took off in earnest. Almost overnight, the Internet—and Network Solutions—became ground zero for a horde of politicians looking to exert their influence and put their stamp on whatever events might transpire and decisions might be made.

This sharply increased interest in the business of Network Solutions on the part of politicians, and government functionaries required the leadership teams of Network Solutions and SAIC to pay particularly close attention to an onslaught of inquiries, investigations, and initiatives coming from Capitol Hill, the White House, and numerous federal agencies on an almost daily basis. I can personally attest that our talented legal and government relations teams were focused on this aspect of our business for most of the five years that we owned NSI in whole or in part.

While we were highly successful in fending off attempts to break our hold on the domain name registry and to dilute our push to commercialize the web, it came at a high cost to our business. It was my personal desire that SAIC conduct the important work it did for its government and commercial customers with a minimum of fanfare and largely out of the limelight. Internally, within the company, we were not shy about trumpeting our successes. However, externally, to the outside world as a whole, we remained modest and quietly successful.

As politicians turned up the heat on SAIC's involvement with Network Solutions, our board reacted to the pressure, beginning to push for the divestment of NSI. We had become quite effective at staying a step ahead of the politicians—anticipating their next moves and meeting or countering them as necessary. Regardless, the spotlight that politicians shined on NSI and its parent SAIC continued to burn bright—further increasing the desire of SAIC's board to sell NSI and remove the source of all this unwanted attention.

In this part, we will examine the sources and reasons behind the intense scrutiny that NSI and SAIC endured as the Internet and the web became commercially important tools for both business and personal use. The lessons we learned are just as applicable to technology businesses today as they were to us in the 1990s. While the technologies may have changed, the nature of politics has not.

12

THE WAR FOR CONTROL OF INTERNET REGISTRIES

Electronic commerce faces significant challenges where it intersects with existing regulatory schemes. We should not assume, for example, that the regulatory frameworks established over the past sixty years for telecommunications, radio, and television fit the Internet. Regulation should be imposed only as a necessary means to achieve an important goal on which there is a broad consensus. Existing laws and regulations that may hinder electronic commerce should be reviewed and revised or eliminated to reflect the needs of the new electronic age.

— President William J. Clinton & Vice President Albert Gore, Jr.[204]

Pressure within the Internet community against the status of Network Solutions, Inc., as the custodian of Internet (IP) addresses and the organization solely responsible for Internet domain name registry continued to grow in the years after we acquired the company. Even absent the fact that NSI was now charging for domain names, many within the community felt that too much power was being vested in this small, Herndon, Virginia-based company and that something should be done about it. As a result, the federal government found itself dealing with a rising tide of discontent over the National Science Foundation cooperative agreement with NSI.

Concern about NSI's monopoly position wasn't just directed at its position as the domain name registrar, it was also directed at one of its other key tasks under the National Science Foundation cooperative agreement: to make network number assignments—that is, to assign IP addresses to individuals and organizations requesting them in North America, Central America, South America, and the Caribbean.

Figure 12-1. Photo of Mike Daniels at NSI HQ briefing Senator Jeff Bingaman, D-NM, Senator Chuck Robb, D-VA, and Senator Tom Daschle, D-SD (Senate Majority Leader), 1999. Reproduced with permission from Michael A. Daniels.

The InterNIC—administered by Network Solutions under its cooperative agreement—was the first of three regional Internet registries (RIRs) established by IANA to distribute IP addresses to individuals and organizations that needed them. The InterNIC—the registry of last resort—serviced the North American region (which at the time also included Central and South America, the Caribbean, and parts of Africa); RIPE NCC serviced the European region: and APNIC serviced the Asia-Pacific region. These RIRs were established when it became clear that one central registry could not effectively serve the global needs of the fast-growing Internet.

There were three goals in the distribution of IP addresses:

1. Conservation: Fair distribution of globally unique Internet address space according to the operational needs of the end-users and Internet Service Providers operating networks using this address

space. Prevention of stockpiling in order to maximize the lifetime of the Internet address space.

2. Routability: Distribution of globally unique Internet addresses in a hierarchical manner, permitting the routing scalability of the addresses. This scalability is necessary to ensure proper operation of Internet routing, although it must be stressed that routability is in no way guaranteed with the allocation or assignment of IPv4 addresses.

3. Registration: Provision of a public registry documenting address space allocation and assignment. This is necessary to ensure uniqueness and to provide information for Internet troubleshooting at all levels.[205]

To accomplish these goals, a three-tier system collectively called the Internet Registry (IR) was developed by the Regional Registries to implement the guidelines developed by the IANA. As described in RFC 2050, the three tiers of this new Internet Registry included:

IANA

> The Internet Assigned Numbers Authority has authority over all number spaces used in the Internet. This includes Internet Address Space. IANA allocates parts of the Internet address space to regional IRs according to its established needs.

Regional IRs

> Regional IRs operate in large geopolitical regions such as continents. Currently there are three regional IRs established: InterNIC serving North America, RIPE NCC serving Europe, and AP-NIC serving the Asian Pacific region. Since this does not cover all areas, regional IRs also serve areas around its core service areas. It is expected that the number of regional IRs will remain relatively small. Service areas will be of continental dimensions.

> Regional IRs are established under the authority of the IANA. This requires consensus within the Internet community of the region.

A consensus of Internet Service Providers in that region may be necessary to fulfill that role.

The specific duties of the regional IRs include coordination and representation of all local IRs in its respective regions.

Local IRs

Local IRs are established under the authority of the regional IR and IANA. These local registries have the same role and responsibility as the regional registries within its designated geographical areas. These areas are usually of national dimensions.[206]

The idea of delegating IP address assignment and registration on an international basis was first put forward by Vint Cerf in RFC 1174—Identifier Assignment and Connected Status—in August 1990. As recommended in this RFC, "With the rapid escalation of the number of networks in the Internet and its concurrent internationalization, it is timely to consider further delegation of assignment and registration authority on an international basis."[207] This idea was further developed and refined within RFC 1366—published in October 1992:

The major reason to distribute the registration function is that the Internet serves a more diverse global population than it did at its inception. This means that registries, which are located in distinct geographic areas, may be better able to serve the local community in terms of language and local customs. While there appears to be wide support for the concept of distribution of the registration function, it is important to define how the candidate delegated registries will be chosen and from which geographic areas. Based on the growth and the maturity of the Internet in Europe, Central/South America and the Pacific Rim areas, it is desirable to consider delegating the registration function to an organization in each of those geographic areas.[208]

InterNIC served as the first RIR, and the second RIR—the European-region RIPE (Réseaux IP Européens) NCC—was established in

Amsterdam in April 1992, providing services to Europe, the Middle East, Africa, and parts of Central Asia. The third RIR—APNIC (Asia-Pacific Network Information Center)—providing service to the Asian region, was established in 1992 by the Asia Pacific Coordinating Committee for Intercontinental Research Networks (APCCIRN) and the Asia Pacific Engineering and Planning Group (APEPG). These two groups were later amalgamated and renamed the Asia Pacific Networking Group (APNG).[209]

Today there are a total of five RIRs, with LACNIC (the Latin American and Caribbean Internet Addresses Registry, headquartered in Montevideo, Uruguay) officially taking over responsibility for IP address allocation in Central and South America and most of the non-English-speaking islands in the Caribbean region in 2002, and AfriNIC (the African Network Information Center, headquartered in Ebene City, Mauritius) officially taking over responsibility for IP address allocation in Africa in April 2005.[210]

The growing concern about the allocation of IP addresses within the region serviced by the InterNIC RIR came from the large telephone companies such as AT&T and MCI, which handled Internet traffic, and the emerging Internet Service Providers (ISPs), which provided Internet access to individuals and businesses. These telephone companies and ISPs required large blocks of IP addresses to enable them to service customer needs. The fact that the decision of whether to issue these blocks of IP addresses was in the hands of a small organization that might slow down or even reject the telephone company's or ISP's request made them nervous. And if that wasn't enough, the small organization—Network Solutions, Inc.—was, unlike RIPE NCC and APNIC (which were nonprofits), a private, for-profit business. What if, for competitive reasons, NSI decided to sit on an IP address allocation request? If the telcos and ISPs didn't have IP numbers to give out to their customers, then they would effectively be out of the Internet business.

And, without even trying, Network Solutions was increasingly becoming a bottleneck for the distribution of IP addresses to these large and politically influential companies. RFC 2050—Internet Registry IP Allocation Guidelines, published in November 1996—required the RIRs to go through a specific and potentially time-consuming process before handing out IP addresses. As described by RFC 2050:

Before a registry makes an assignment, it must examine each address space request in terms of the requesting organization's networking plans. These plans should be documented, and the following information should be included:

1. subnetting plans, including subnet masks and number of hosts on each subnet for at least one year
2. a description of the network topology
3. a description of the network routing plans, including the routing protocols to be used as well as any limitations

The subnetting plans should include:

a. a tabular listing of all subnets on the network
b. its associated subnet masks
c. the estimated number of hosts
d. a brief descriptive remark regarding the subnet

If subnetting is not being used, an explanation why it cannot be implemented is required. Care must be taken to ensure that the host and subnet estimates correspond to realistic requirements and are not based on administrative convenience.[211]

Pressure grew on the federal government, and specifically on the National Science Foundation, to take action to split NSI's IP number allocation function from the cooperative agreement. Eventually, NSI was encouraged by NSF to separate the management of domain name registration from the management of IP numbers. As a direct result, the 501(c)(6) nonprofit American Registry for Internet Numbers, Ltd.—ARIN—was formed and funded by NSI in April 1997, with a membership structure open to all interested entities and individuals, including ISPs, end-user organizations, corporations, universities, and others.[212]

According to George Strawn, the impetus for the creation of ARIN began a couple of years before the split. Says Strawn, "From my point of view, the decision to spin off ARIN came early, late 1995 or early 1996, and it resulted from one of the workshops NSF held with the traditional Internet community after imposing registration fees—to mend fences. It was pointed out that no other entity was in charge of both names and numbers. So to make the separation in the US would bring us into conformity with the other regions of the world."[213]

I was personally against the idea of spinning off NSI's control of the management of IP numbers because I recognized the value that IP numbers could play in the future, but I was eventually convinced that it was in the best interests of both NSI and SAIC.

Phil Sbarbaro—a member of the law firm Hanson and Molloy in Washington, DC, who served as NSI's outside general counsel for a number of years, and before that as counsel at the US Department of Justice and the US Securities and Exchange Commission—recalls a conference call with me on the topic. Says Sbarbaro, "There was a conference call with Dr. Beyster. Everyone called in from all around the world; all different time zones. I think it was 4:00 a.m. for us. I explained that we had to give up IP numbers. The assembled conference call was deathly silent because they knew what was coming far better than I did. Dr. Beyster asked me this question: 'Why are you dismantling my company?' That was my first direct confrontation with Dr. Beyster, and no one came to my defense on that call. I had to explain that the government is going to take it anyway. And if the government takes it, it will be far more bloody and far more public. We give the numbers away without charge; we have no contract with the entities to whom we give them."[214]

NSI put up the money to start up ARIN, and we transferred the entire IP registry team—led by Kim Hubbard—from NSI into this new organization. Kim became ARIN's president and one of the founding directors. Phil Sbarbaro was the first chairman of the board and also served as founding director along with Don Telage; both helped set up the organization. Phil focused on getting ARIN's articles of incorporation and 501(c)(6) paperwork done, and Don focused on securing office space, buying furniture and other equipment, hiring employees, and working out the legal complications (for example, with stock options) for NSI employees who were transferred to the new organization. We fronted ARIN and guaranteed its operations for a full year, until it gained members and was able to operate independently.[215] Don Telage also searched for and convinced industry participants to join the board of trustees to give it credibility for charging for IP numbers.[216]

According to ARIN's articles of incorporation, the organization's reason for being formed included the following purposes:

1. to increase and diffuse knowledge to the general public about the Internet in its broadest sense;
2. to educate industry and the Internet community in order to further their technical understanding of the Internet;
3. to secure united action and to represent the Internet community nationally and internationally;
4. to manage and help conserve scarce Internet protocol resources, and to educate Internet protocol users on how to efficiently utilize these scarce resources as a service to the entire Internet community;
5. to do all and everything necessary to enhance the growth of the Internet and the prospects for competition among Internet Service Providers by encouraging the exploration and implementation of solutions to Internet Protocol number scarcity issues;
6. to encourage the exploration of new addressing and routing technologies that reduce or eliminate the costs or in some cases the need for renumbering when an Internet Service Provider or end user changes to a new Internet Service Provider; and, when such alternatives are developed, to work with its members to facilitate the assignment of portable addresses and/or the elimination of the cost of Internet Protocol renumbering;
7. to encourage allocation policy changes for Internet Service Providers in order to enhance competition by providing mobility of Internet Service Providers among upstream Internet Service Providers when it is generally agreed that the technology is available for portable addressing;
8. to manage the allocation and registration of Internet resources;
9. to promote and facilitate the expansion, development, and

growth of the infrastructure of the Internet for the general public and members by any means consistent with the public interest through other activities, including, but not limited to, publications, meetings, conferences, training, educational seminars, and the issuance of grants and other financial support to educational institutions, foundations, and other organizations exclusively for educational, charitable, and scientific purposes.[217]

According to Kim Hubbard, "After NSI began charging for DNS registration and the discussion on DNS registration competition began, there was concern voiced among some in the industry that IP registration might somehow be affected. DNS registration competition discussions at one point became quite hostile, and the last thing anyone wanted was IP registration to be pulled into the political morass that DNS had become. There was concern that while it was plausible to make DNS registration a financially competitive function, the same could not be said for IP registration. For various technical reasons, it was not viable for companies to compete on a monetary basis to allocate IP numbers. Continued Hubbard, "it became evident that the best way to handle the situation was to separate the two registration processes in every way possible." [218]

The creation of ARIN put the IP address system in a stable position, and it was enough to remove some of the pressure that the federal government felt because of Network Solutions. For many, though, it wasn't enough. There were still significant concerns about NSI's status as the sole Internet domain name registry. And rather than going away, these concerns continued to build.

Ira Magaziner—President Clinton's senior advisor on policy—was given responsibility for looking into NSI's monopoly status and taking action to dismantle it. It was very much in the interest of NSI and SAIC to find a win-win solution that would look out for our own business interests, protect the integrity of the Internet, and satisfy the many stakeholders. These stakeholders included the US government because it wanted to protect this future platform for business. The international community had a stake in it because it was going to be a worldwide web, and the Europeans had already made contributions to its development, specifically CERN's development of the web. The business community had a vested interest, and it wanted to be secure. And there were the keepers of the Internet—including the Internet Society, Jon Postel, and others who considered the Internet their own creation.

Jon Postel decided that it was time to show these other stakeholders just exactly who held the keys to the domain name system.

13

THE E-MAIL MESSAGE HEARD 'ROUND THE WORLD

As the Internet develops there are transitions in the management arrangements. The time has come to take a small step in one of those transitions.

— Jon Postel, January 28, 1998[219]

Jon Postel was unhappy with the monopolistic power that the for-profit company Network Solutions had gained over the Internet, and when it came to all things Internet, Jon Postel was one person you wanted on your side, not working against you. In their book about the origins of the Internet, *Where Wizards Stay Up Late*, authors Katie Hafner and Matthew Lyon explain that Postel thought that the Internet belonged to the people—that it was the modern-day equivalent of the ancient idea of the public commons—and that using it as a profit-making enterprise was anathema to the original intent of its founders as well as its core culture:

> He worked quietly for years as keeper of the RFCs and final arbiter in technical matters when consensus couldn't be reached. Postel believed that decisions he had made in the course of his work over the years had been for the good of the community and that starting a company to profit from those activities would have amounted to a violation of public trust.[220]

For most of the history of the DNS up to that time, no one really paid much attention to the root name servers' operators. Mark Kosters maintained a mailing list for them to help enable communication within the group, but that was about it. As the Internet became more important, Jon Postel as head of IANA tried to assert more authority over the root name servers. Finally, on January 28, 1998, Postel sent e-mail messages to eight of the root name server operators—the ones controlling the non–US government roots—directing them to switch the root zone server from Network Solutions' A server (A.ROOT-SERVERS.NET at IP address 198.41.0.4) to a server controlled by IANA: DNSROOT.IANA.ORG at IP address 198.32.1.98. The root name server operators complied with Postel's order—taking the Internet out of NSI's and the US government's control.[221]

Figure 13-1. Photo of Jon Postel. *Photo by* Irene Fertik, USC News Service. Copyright 1994, USC. Permission granted for free use and distribution, conditioned upon inclusion of the above attribution and copyright notice.

Here is the text of Postel's e-mail message to the eight root name server operators:

```
Date: Wed, 28 Jan 1998 17:04:11 -0800
From: postel@ISI.EDU
Subject: root zone secondary service
```

Cc: postel@ISI.EDU, iana@ISI.EDU
The following messages is pgp signed by "iana."

—BEGIN PGP SIGNED MESSAGE—

=====================================
=====================================

Hello.

As the Internet develops there are transitions in
the management arrangements. The time has come to take a
small step in one of those transitions. At some point on
down the road it will be appropriate for the root domain
to be edited and published directly by the IANA.

As a small step in this direction we would like to
have the secondaries for the root domain pull the root zone
(by zone transfer) directly from IANA's own name server.

This is "DNSROOT.IANA.ORG" with address 198.32.1.98.

The data in this root zone will be an exact copy
of the root zone currently available on the A.ROOT-SERV-
ERS.NET machine. There is no change being made at this
time in the policies or procedures for making changes to
the root zone.

This applies to the root zone only. If you provide
secomdary [sic] service for any other zones, including
TLD zones, you should continue to obtain those zones in
the way and from the sources you have been.

 – --jon.

```
Jon Postel
Internet Assigned Numbers Authority
c/o USC - ISI, Suite 1001
4676 Admiralty Way
Marina del Rey, CA 90292-6695

Talk: +1-310-822-1511
Fax: +1-310-823-6714
E-mail: IANA@ISI.EDU


====================================
====================================222
```

Mark Kosters recalls what happened on that fateful day. Says Kosters, "I noticed on the logs one day that a number of the root servers were no longer fetching the root zone from us—they were fetching it from someone else. So I sent one of the root server operators a message saying 'Your machine's misconfigured because it's not getting information from the right source.' He sent back a message that said, 'I got a note from Jon Postel saying that I needed to get the information from some other machine.' At that point, I was like 'Uh-oh—this isn't right.' So I called Jon Postel to talk about his root server redirect. I said, 'Jon, you can't do this. This whole thing has become very political. Changing the configuration of the Internet and how it works right now is not a good idea.' Jon told me he was going to do it anyway because NSI was not IANA. I told him that I thought it was a big mistake."[223]

Long story short, Postel had redirected the large part of the Internet not under control of the federal government and Network Solutions and made it answer only to himself. Not only did this loudly demonstrate to the government, Network Solutions, and the Internet community as a whole that he (and IANA) held the ultimate authority when it came to all things Internet, but it sent the message loud and clear that the people who had designed and built the Internet's architecture and its unique open culture were not happy with the way it was being commercialized by being put in the hands of a for-profit company. The federal government had funded much of the research, development, and infrastructure-building of the Internet, so the United States had its own strong claim of ownership—one that ultimately trumped those people and organizations involved in the network's inception and increasingly rapid growth.

Regardless, not one of the operators of the eight affected root servers had any reason to question the direction they received from Postel—he was head of IANA, after all. According to Gerry Sneeringer, operator of the D root server, "If Jon asks us...we'll do it. He is the authority here."[224] But that didn't mean that some of the operators weren't nervous about making the change. Paul Vixie, who operated the F root server for Internet Software Consortium of Redwood City, California, was said to have taken an extra precaution: he arranged to have his children taken care of in his absence in the event that he was arrested for following Postel's orders.[225]

It didn't take long for word to get out about Postel's action. Kosters immediately called Dave Holtzman to let him know what had happened—that all the root name servers except for the government's NASA, .mil, US Army Ballistics Research Laboratory (BRL), and the one at NSI were now pointing to IANA's root zone server instead of NSI's. Ira Magaziner was at the World Economic Forum in Davos, Switzerland. When US national security officials found out about Postel's redirection of the name servers, they woke up Magaziner at one in the morning to notify him that something odd was going on with the Internet root system. By two o'clock, Magaziner had Postel on the telephone line, along with a supervisor of Postel's at USC for extra measure. According to Magaziner, the conversation went like this:

Magaziner: Jon, what is going on with the Internet Root?

Postel: We were simply conducting a test.

USC official: [Gasp] You were doing what? [tone of disbelief and anger]

Magaziner: Jon, you don't have the legal right to conduct a test. You cannot conduct a test without DARPA's approval. You will be in trouble if you continue this; both you and USC will be liable.

USC official: Hell, we could have lawsuits up the kazoo because of the impact of this on commerce. It could bankrupt the university. Jon, you have to stop this immediately.

Postel: Sorry, I was just doing a test. I didn't mean to do anything wrong.

Magaziner: We don't want to cause you any trouble. Put things back as they were and we'll agree to call this a test.[226]

According to Mark Kosters, "The United States government came back to Jon Postel and told him in no uncertain terms that he needed to immediately change the name servers back. They even called in the US Marshals to force him to make the change back, and he finally did. He sent a note to all the root server operators that said, in essence, 'Please change all your root name servers back to the one that's operated by Network Solutions, and thank you for working on this test with us.'"[227]

This put control of the root zone server—and, therefore the Internet—firmly in the hands of the US government, where it has resided ever since.

The idea of a for-profit company being in charge of domain name registration was an ongoing challenge, both for NSI and SAIC. The old-school Internet folks were not free-market oriented. And then at the same time, Network Solutions was viewed as a monopoly—we were for some time the only domain name registry and, at that time, registrar. In fact, there wasn't even a distinction of the two roles. People saw this great growth in NSI's business, and they naturally wanted competition—they wanted to play in that world and to have a chance to get a piece of the action. As that effort began to evolve, Network Solutions fell under tremendous criticism just because of the fact that it was the only one handing out domain names—the 800-pound gorilla.

The government couldn't ignore for long the fact that NSI attracted all this negative publicity. According to Jay Killeen, senior vice president for government affairs at SAIC, "I think that the Commerce Department, especially later on, and some Democrats in Congress really took umbrage to the fact that we had made such a windfall on Network Solutions. It was too much; therefore, there had to be something wrong. So they started to investigate—looking into the contractual relationship with the National Science Foundation and looking into our business dealings. They were looking for a smoking gun—that something had to be wrong with this arrangement, and that's why the government should take this back. There was a clamor for hearings."[228]

Network Solutions' leadership team did make many attempts to defuse the negative feelings toward the company. Specifically, Chuck Gomes started participating in the domain policy list, an often raucous Internet public forum hosted by NSI for the discussion of domain policy and intellectual property issues. Chuck was the first employee from Network Solutions to actively start participating in that list, and it was quite a challenge for him due to the frequent bashing of NSI by participants. Regardless, over time, Gomes was able to develop some trust with people on the list and build productive and long-term working relationships.

The domain policy list went on for years, but NSI finally had to shut it down because legal counsel said it was too vulnerable—the company was under huge amounts of criticism. The International Ad Hoc Committee (IAHC) and the Internet Society (ISOC) began to develop their own ideas of how the domain registration business should be managed, and the ideas they were coming up with were very top down. They envisioned controlling Network Solutions and anybody else that was a registry. And so we at SAIC—along with NSI's management team—began getting actively involved in the policy around the Internet to advocate for a more free-market-oriented system. And we had some big battles in that regard.

These battles led to fundamental changes in NSI's role in the administration of the domain name system.

14

A POSTEL MEMO, AND GREEN AND WHITE PAPERS

The system for registering second-level domain names and the management of the TLD registries should become competitive and market-driven.

— NTIA Green Paper, January 30, 1998[229]

Although the creation of ARIN in December 1997 took NSI's administration of IP addresses off the table—defusing one flashpoint for Internet purists—the war over control of the Internet was just getting started. And the primary target in this war was NSI's monopoly over administration of the domain name system.

In May 1996, Jon Postel fired the opening salvo in this new theater of the rapidly escalating DNS war when he issued a draft memo titled, "New Registries and the Delegation of International Top Level Domains." In this memo (which he describes as "a total rip off of a draft by Randy Bush, combined with substantial inclusion of material from a draft by Karl Denninger"), Postel outlines a set of policies and procedures to allow open competition in domain registration in the iTLDs (international top-level domains) and to provide the IANA with a legal and financial umbrella that would support its ongoing operations.[230]

Speaking for himself—but also for many Internet purists and long-standing members of the Internet community who, like him, were unhappy with the creeping commercialization of the Internet—Postel stated in his memo:

There is a perceived need to open the market in commercial iTLDs to allow competition, differentiation, and change, and yet maintain some control to manage the Domain Name System operation.

The current situation with regards to these domain spaces, and the inherent perceived value of being registered under a single top level domain (.COM) is undesirable and should be changed.

Open, free-market competition has proven itself in other areas of the provisioning of related services (ISPs, NSPs, telephone companies) and appears applicable to this situation.

It is considered undesirable to have enormous numbers (100,000+) of top-level domains for administrative reasons and the unreasonable burden such would place on organizations such as the IANA.

It is not, however, undesirable to have diversity in the top-level domain space, and in fact, positive market forces dictate that this diversity, obtained through free competition, is the best means available to insure quality service to end-users and customers.[231]

The mechanism by which Postel and IANA would wrench sole control of the DNS away from Network Solutions was by the creation of an entirely new set of top-level domains, along with new registries to operate each of them. Oversight of this new system would be accomplished by IANA and a new organization that Postel labeled "the ad hoc committee." Postel proposed that the ad hoc committee "consist of seven members appointed by the IANA (3), the IETF (2), and the ISOC (2)."[232] Although Postel specifically chose not to directly take on the longer-term issue of the management and charter of the then-current iTLDs (.com, .net, and .org, administered by Network Solutions), or the specialized TLDs (.edu, .gov, .mil, .int, and .arpa, administered by the federal government) in his memo, the message was clear: competition was coming.

Postel estimated that in the first year of his proposed plan, up to fifty new domain name registries would be chartered to compete with Network Solutions, with no more than two-thirds chartered within the same country. Each new registry would choose up to three new iTLD names to administer under its charter—resulting in a potential total of up to 150 new top-level domains in the first year—and these registries would have the exclusive right to administer these domain names for a period of time limited to five years. After the end of the first year of his proposed plan, Postel anticipated that during the next four years, approximately thirty new iTLDs would be created each year, along with approximately ten new registries chartered each year to administer the new iTLDs. In each year thereafter, additional iTLDs would be created and new registries chartered as needed.[233]

At their June 1996 annual meeting in Montreal, the board of trustees of the Internet Society (ISOC) voted to adopt Postel's draft memo in principle while encouraging its further refinement.[234] This action breathed life into Postel's proposal, leading to the formation of the Internet Ad Hoc Committee (IAHC) by IANA and ISOC in October 1996.

According to Don Heath, then president and CEO of the Internet Society, the goal of the IAHC was to "undertake defining, investigating, and resolving issues resulting from current international debate over a proposal to establish global registries and additional Top Level Domain names (iTLDs)."[235] The founding board of the IAHC was comprised of representatives from ISOC, IANA, the Internet Architecture Board (IAB), the International Telecommunication Union (ITU), the World Intellectual Property Organization (WIPO), and the International Trademark Association (ITA).

On February 4, 1997, the IAHC released a report that proposed a Generic Top Level Domain Memorandum of Understanding (gTLD-MoU). This MoU would enable the creation of seven new generic top-level domain names (gTLDs) that would be operated by new (presumably non-Network Solutions) registries. The IAHC proposed that these registries would be supervised by a new organization: the Council of Registrars (CORE).

These new gTLDs included:

.firm   for businesses, or firms
.store   for businesses offering goods to purchase
.web   for entities emphasizing activities related to the World Wide Web
.arts   for entities emphasizing cultural and entertainment activities

.rec   for entities emphasizing recreation/entertainment activities

.info   for entities providing information services

.nom   for those wishing individual or personal nomenclature, i.e., a personal nom de plume[236]

While the IAHC proposal gained some support, and the proposed gTLD-MoU was ultimately drafted (with .shop substituted for .store),[237] the proposal largely fell flat within the Internet community. The proposed timeline for development and implementation of about one hundred days was considered too ambitious by many, and others felt that the proposal didn't really address and resolve the issues of competition within the domain name space. A lack of unity over the IAHC's proposal resulted in no action being taken toward creation of Jon Postel's proposed new set of top-level domains and the chartering of the new registries to operate them. In 1997—soon after the IAHC issued its report—the organization dissolved.[238] (See Chapter 18 for a discussion of ICANN's current plan to introduce new gTLDs, including new ASCII and internationalized domain name (IDN) top-level domains.)

But this wasn't the last attempt to pry Network Solutions loose from its monopoly position. While the Internet old guard may have lost this particular battle, the war was far from over.

On January 30, 1998 (coincidentally, just two days after Jon Postel ordered the redirection of eight of the Internet's root name servers from NSI to IANA), after a series of Congressional hearings, the US government issued a discussion draft. Brought forth by the Department of Commerce and the National Telecommunications and Information Administration (NTIA)—an agency in the US Department of Commerce that serves as the executive branch agency principally responsible for advising the president on telecommunications and information policies—the draft was titled "A Proposal to Improve Technical Management of Internet Names and Addresses."

This document—which became known as the Green Paper and which was drafted under the supervision of Ira Magaziner—followed through on two separate federal government initiatives. The first was the Framework for Global Electronic Commerce, in which President Clinton on July 1, 1997, directed the

Secretary of Commerce to privatize, increase competition in, and promote international participation in the domain name system. The second was the July 2, 1997, Department of Commerce Request for Comments (RFC) on DNS administration, issued on behalf of an inter-agency working group previously formed to explore the appropriate future role of the US government in the DNS. The RFC solicited public input on issues relating to the overall framework of the DNS system, the creation of new top-level domains, policies for registrars, and trademark issues. During the comment period, more than 430 comments were received, amounting to some 1500 pages.[239]

The Green Paper posited four principles for a new domain name system:

1. Stability. The US government should end its role in the Internet number and name address systems in a responsible manner. This means, above all else, ensuring the stability of the Internet. The Internet functions well today, but its current technical management is probably not viable over the long term. We should not wait for it to break down before acting. Yet, we should not move so quickly, or depart so radically from the existing structures, that we disrupt the functioning of the Internet. The introduction of a new system should not disrupt current operations or create competing root systems.

2. Competition. The Internet succeeds in great measure because it is a decentralized system that encourages innovation and maximizes individual freedom. Where possible, market mechanisms that support competition and consumer choice should drive the technical management of the Internet because they will promote innovation, preserve diversity, and enhance user choice and satisfaction.

3. Private, Bottom-Up Coordination. Certain technical management functions require coordination. In these cases, responsible, private-sector action is preferable to government control. A private coordinating process is likely to be more flexible than government and to move rapidly enough to meet the changing needs of the Internet and of Internet users. The private process should, as far as possible, reflect the bottom-up governance that has characterized development of the Internet to date.

4. Representation. Technical management of the Internet should reflect the diversity of its users and their needs. Mechanisms should be established to ensure international input in decision making.[240]

At the root of the government's proposed new system was the transfer of existing IANA functions, the root system, and the appropriate databases to a new not-for-profit corporation (what became known as the "NewCo"). It was anticipated by the government that the transition would commence as soon as possible, with operational responsibility moved to the new entity by September 30, 1998—the end date of NSI's cooperative agreement with the National Science Foundation. In essence, if the recommendations of the Green Paper were enacted as proposed, then on October 1, 1998, Network Solutions would be removed from its lucrative domain name business and replaced by a nonprofit organization.

According to the Green Paper, this new nonprofit corporation would operate as a private entity for the benefit of the Internet as a whole. The new corporation would have authority

1. to set policy for and direct the allocation of number blocks to regional number registries for the assignment of Internet addresses;
2. to oversee the operation of an authoritative root server system;
3. to oversee policy for determining, based on objective criteria clearly established in the new organization's charter, the circumstances under which new top-level domains are added to the root system; and
4. to coordinate the development of other technical protocol parameters as needed to maintain universal connectivity on the Internet.[241]

Although the Green Paper addressed a number of other pressing issues within the Internet community—including the creation of new gTLDs, domain name registries, and registrars (following through on the ideas of Jon Postel and the now-defunct IAHC); the ongoing problem with domain names and trademarks; and the status of the National Science Foundation's Internet Intellectual Infrastructure Fund—the proposed termination of NSI's cooperative agreement was the item of greatest interest to NSI's leadership team, and to me personally.

According to the Green Paper, the US government would ramp down the NSI cooperative agreement and phase it out by the end of September 1998. It was proposed that the ramp-down agreement with NSI would reflect the following terms and conditions designed to promote competition in the domain name space:

1. NSI will effectively separate and maintain a clear division between its current registry business and its current registrar business. NSI will continue to operate .com, .net and .org but on a fully shared-registry basis; it will shift operation of .edu to a not-for-profit entity. The registry will treat all registrars on a nondiscriminatory basis and will price registry services according to an agreed upon formula for a period of time.
2. As part of the transition to a fully shared-registry system, NSI will develop (or license) and implement the technical capability to share the registration of its top-level domains with any registrar so that any registrar can register domain names there in [sic] as soon as possible, by a date certain to be agreed upon.
3. NSI will give the U.S. government a copy and documentation of all the data, software, and appropriate licenses to other intellectual property generated under the cooperative agreement, for use by the new corporation for the benefit of the Internet.
4. NSI will turn over control of the "A" root server and the management of the root server system when instructed to do so by the U.S. government.
5. NSI will agree to meet the requirements for registries and registrars set out in Appendix 1.

On June 5, 1998, the Department of Commerce and NTIA issued a statement of policy—"Management of Internet Names of Addresses," Docket Number: 980212036-8146-02—which restates the proposals of the Green Paper and summarizes comments received by the government in response to these proposals. This statement of policy became known as the White Paper. Regarding the NSI cooperative agreement, the White Paper had this to say:

> **Comments:** Many commenters expressed concern about continued administration of key gTLDs by NSI. They argued that this would give NSI an unfair advantage in the marketplace and allow NSI to leverage economies of scale across their gTLD operations. Some commenters also believe the Green Paper approach would have entrenched and institution-alized NSI's dominant market position over the key domain name going forward. Further, many commenters expressed doubt that a level playing field between NSI and the new registry market entrants could emerge if NSI retained control over .com, .net, and .org.

> **Response:** The cooperative agreement between NSI and the U.S. government is currently in its ramp-down period. The U.S. government and NSI will shortly commence discussions about the terms and conditions governing the ramp-down of the cooperative agreement. Through these discussions, the U.S. government expects NSI to agree to take specific actions, including commitments as to pricing and equal access, designed to permit the development of competition in domain name registration and to approximate what would be expected in the presence of marketplace competition. The U.S. government expects NSI to agree to act in a man-ner consistent with this policy statement, including recognizing the role of the new corporation to establish and implement DNS policy and to establish terms (including licensing terms) applicable to new and existing gTLD registries under which registries, registrars, and gTLDs are permit-ted to operate. Further, the U.S. government expects NSI to agree to make available on an ongoing basis appropriate databases, software, docu-mentation thereof, technical expertise, and other intellectual property for DNS management and shared registration of domain names.

But while the proverbial writing seemed to be on the wall, Network Solu-tions was not about to take this proposal lying down. The stakes were too high, there was too much money in play, and we were determined that NSI's exclusive Internet franchise would not be ended prematurely. This meant engaging in a series of discussions with the government and key Internet stakeholders—ultimately working out a compromise that everyone could live with, if not be happy about.

According to Jay Killeen, who was very much involved in educating members of Congress about NSI's critical role in the stability of the Internet, "We perceived there would be a problem down the road—or at least a lot of questions, and a lot of confusion—so we started talking to a lot of members of Congress, both House and Senate, and both Republicans and Democrats. We established ourselves as subject matter experts on the Internet—we built up credibility. We were able to have members ask us questions, and we were able to provide answers for things that didn't even pertain to Network Solutions. So we became one of many go-to experts on this phenomenon, which was the Internet, and we built up some credibility through that."[242]

And as the government slowly maneuvered its way to a showdown with NSI, the company needed all the credibility it could get.

15

ICANN AND THE BEGINNING OF THE END...OR NOT

We over-regulated some stuff, and we under-regulated other stuff. We got every-body to despise us. We didn't listen. And, unfortunately, it hasn't gotten much better since.

— Esther Dyson, founding chairperson, ICANN[243]

The publication of the government's White Paper gave momentum—and the blessing of the Clinton administration—to the idea of creating a nonprofit organization that would take over Internet policy setting while creating competition for and oversight over Network Solutions. In a press conference held in conjunction with the release of the White Paper, Department of Commerce spokesperson Becky Burr said:

> We are looking for a globally and functionally representative organization, operated on the basis of sound and transparent processes that protect against capture by self-interested factions, and that provides robust, professional management. The new entity's processes need to be fair, open, and pro-competitive. And the new entity needs to have a mechanism for evolving to reflect changes in the constituency of Internet stakeholders.[244]

In the months that followed, numerous forums, workshops, and global online discussions were initiated to discuss the formation of a nonprofit NewCo to administer the DNS. Individuals and organizations from fifty different countries participated in the discussions about the formation of NewCo, and five drafts of a legal framework for the formation of NewCo were circulated widely on the Internet.[245] In 1998 alone, four international meetings were held in response to the White paper—the first in Reston, Virginia; the second in Geneva; the third in Singapore; and the fourth in Buenos Aires.

Figure 15-1. Photograph of President Bill Clinton with Mike Daniels (right), 2000. Reproduced with permission from Larry Glenn, Photo-Op, Inc. and Michael A. Daniels.

On October 2, 1998, Jon Postel was finally able to pound a lasting stake into the heart of NSI's hold on the domain name system. On this day, Postel—as director of IANA—transmitted a draft of NewCo bylaws and articles of incorporation to William Daley, the then Secretary of Commerce. In the articles of incorporation, Postel gave NewCo a new name: the Internet Corporation for Assigned Names and Numbers, or ICANN for short. Jon Postel died soon afterward, on October 16, 1998, of complications from surgery to repair a replacement heart valve.[246] The ICANN board of directors held its first meeting—in Cambridge, Massachusetts—on November 14, 1998, and on

November 25, 1998, ICANN and the Department of Commerce signed a memorandum of understanding that officially designated ICANN as the NewCo.[247]

From that point on, Network Solutions and ICANN were at war with one another, and the birth and early days of ICANN were rocky to say the least. Esther Dyson, ICANN's founding chairperson, was there close to the beginning, and she was very much aware of how the various Internet stakeholders felt about the mission of this soon-to-be created organization. Says Dyson, "The word got out that some kind of body was going to be created, and the Congresspeople who got wind of it thought it was giving up an American birthright to a bunch of dirty foreigners. The commercial people said, 'The thing works; don't mess with it.' And the human rights activists said, 'This is going to be a tool of corporate interests.' The people outside the US thought it was just some kind of weird American trick to pretend to give it away. So nobody trusted anybody."[248]

In the meantime, the clock had been relentlessly ticking away on the term of NSI's cooperative agreement with the NSF. The agreement end date of September 30, 1998, came and went, and Network Solutions was still in charge of DNS administration. Behind the scenes, NSI's management team was earnestly negotiating the terms of an amendment to the cooperative agreement that would allow the company to hold on to as much of the DNS-related business as possible, while granting the government the concessions it needed to fulfill the terms of the White Paper and ICANN's charter.

The negotiations were quite contentious. Recalls Mark Kosters, "I was not directly involved, but Dave Holtzman sat down with me and said 'This is not going well. How can I fix this?' And I said, 'Well we may want to consider separating out the registrar and the registry functions, where the registrar has most of the company information. That could be used for multiple other purposes and that's probably where the money's going to be. And then the registry would be very thin, and all it would have is the domain name, the name servers associated with it, and the registrar of record—who registered it. You could then have multiple registrars come in and register these things.' Frankly, it wasn't my idea—it was actually proposed back in the 1992 timeframe for the InterNIC. The National Science Foundation thought the idea was too risky at the time, and that's why they stayed with Network Solutions."[249]

Dave Holtzman presented the idea to the Department of Commerce, and the negotiations went well from that point forward. And that started the registrar-registry split that we have today.[250]

On October 7, 1998, the Department of Commerce issued Amendment No. 11 to NSI's cooperative agreement with the National Science Foundation. Among other things, this amendment extended the end date of the cooperative agreement to September 30, 2000, allowing Network Solutions to continue to operate the primary root name server while requiring NSI to recognize NewCo [ICANN] and to cooperate with the organization in the orderly transition of DNS responsibilities.[251]

In addition, the amendment set a timetable for development of a Shared Registration System that would enable other "Accredited Registrars" to offer domain name registration services to the public. The timetable in the amendment was as follows:

1. By November 1, 1998, NSI shall provide functional and interface specifications for the Shared Registration System and a milestone schedule for its development and implementation.
2. By December 1, 1998, NSI shall create a focused input technical advisory group consisting of not more than ten individuals designated by NewCo to comment on the design of and participate in testing of the Shared Registration System.
3. By March 31, 1999, NSI will establish a test bed supporting actual registrations in .com, .net and .org by five registrars accredited by NewCo (Accredited Registrars). (Phase 1)
4. By June 1, 1999, the Shared Registration System will be deployed by NSI and available to support multiple licensed Accredited Registrars offering registration services within the gTLDs for which NSI now acts as a registry. (Phase 2)
5. By October 1, 1999, NSI will have completed reengineering of NSI's registry/registrar interface and back end systems so as to assure that NSI, acting as registry, shall give all licensed Accredited Registrars (including NSI acting as registrar) equivalent access ("equal access") to registry services through the Shared Registration System. (Phase 3). The functional and interface specifications of the Shared Registration System shall

describe a protocol and associated software able to: (1) provide security and authentication protocols and procedures for requests from registrars; and (2) permit second level domain name holders to change registrars within the same registry without changing domain names.

NSI agrees to license the Shared Registration System protocol, associated documentation, and reference implementation to Accredited Registrars, on reasonable terms and conditions approved by the USG, such approval not to be unreasonably withheld, that are designed to promote the development of robust competition for the provisions of registrar services.

The idea to split apart the registry and registrar functions was originally NSI's, not the government's. As former Network Solutions CEO Gabe Battista explains in an interview, "We came up with the concept, consistent with the deregulation in the telcom industry where long distance service was competitive, but local access was provided by the Bell Companies. After kicking it around among ourselves, and then having a meeting with Dr. Beyster and others, we decided that there should be a registry that houses all the names—that is totally safe and secure, and which gets some minimal amount of money from managing that infrastructure. And then for all the people that were worried how much you charge for registering, we would have individual registries. And anybody could become a registrar, but there could only be one registry."[252]

Once it was clear that NSI could not maintain the sole source of domain name registrations in the major .com, .net, and .edu registries, there was a big strategy meeting at Network Solutions about what to do next. Phil Sbarbaro recalls that at this meeting, the idea of splitting the registry and registrar functions was first voiced. Says Sbarbaro, "I had discussed it before the meeting, of all things, with my wife who is an airline tariff specialist at the World Bank. I explained that all of the registrations must emanate from one source, but I don't know how to split that function up. She said, 'We do it every day.' I asked, 'How?' Her response was, 'Atlas and Sabre. There are only a limited number of airline seats. The airline must issue the ticket, but travel agents along with the airlines sell them. Travel agents hook into the main database. If a seat is taken, it's taken. If it isn't, it's available.' The NSI meeting was at about noon, and I was getting tired of listening to all these people debate. We're running out of time. I took the chalk and I started at the far end of the board by the window and worked my way across. 'These functions above the line are the registry. These functions below the line are the registrar. This is the airline above and Atlas below. These are like travel agents. The airline can sell seats and the travel agents can sell seats. Once they are booked, they are booked. I have to go.' As I walked out the door, everybody starts cheering and clapping. Now I don't have time. I'm late for court."[253]

And that is exactly what happened. Network Solutions maintained its position as the Internet registry to manage the domain names, while other companies could sell domain names, competing on price and service. Each time a company sold a domain name, a small amount of the proceeds of that sale would go to Network Solutions as the registry. And although this amount of money was relatively small, when multiplied by the many millions of domain name registrations that were occurring, it was still a significant amount of money.

On April 21, 1999, ICANN announced its selection of thirty-four accredited domain name registrars to compete with NSI. Of these thirty-four organizations, five were chosen to take part in the Shared Registry System test bed required under the terms of the amended cooperative agreement. The five organizations included America Online, CORE (Internet Council of Registrars), France Telecom/Oleane, Melbourne IT, and register.com.[254] On June 7, 1999—within a couple weeks after Jim Rutt was brought in as NSI's new CEO in May 1999, taking over the company reins from Mike Daniels, who had held the position on an interim basis since Gabe Battista quit in December 1998[255]—Network Solutions announced that register.com was the first of these five registrars to successfully register a domain name—ending NSI's monopoly of the DNS.[256]

When it came time to implement the separation of the registry and registrar functions, Dave Holtzman made the decision to implement a thin registry model. According to Holtzman, "A fat registry means that all the personal details of the domain name owners would have been incorporated into a large central repository with relatively thin registrars that would then become sales organizations. The alternative model was the one we chose to use: the thin registry. In the thin registry model, the registry never has access to any the domain name owner's personal information. Instead, it's up to the registrars to maintain the details of the client relationship, and it's up to the registry to maintain a connection with the registrar."[257]

Recalls Holtzman, "I made the decision, as much for business reasons as technical ones. A fat registry would not only be more prone to breakage because it was a single, centralized point of failure, but it didn't easily achieve the objective of creating a multi-company, global trade in domain names. Even more important was the legal/political consideration that a fat registry would have to abide by the sum of all the laws and regulations of most of the world's nations, whereas a thin registry was only subject to American rules, pushing the regional issues out to where they appropriately belonged—to the sovereign nation of any one particular registrar."

Holtzman continues, "This is a fairly significant point. If the fat registry had been implemented—which Pincus, NTIA, and I think NIST wanted—it would have become bloated and quickly mired down in global litigation and regulation, slowing the growth of the then burgeoning Internet."[258]

As it turned out, the key people on the government side—Becky Burr and Andy Pincus in the Commerce Department, and Ira Magaziner at the White House—were furious at the idea that Network Solutions had adopted a thin registry. The government wanted a fat registry.

I suspect, however, that the last thing the Commerce Department wanted to do was to open up the issue for comments and have everybody on the Internet wade in with their opinions. Based on our experience with Network Solutions, we found that if there are a million people on the Internet, you will have a million different opinions. If the Department of Commerce had put the issue out for comments on the Internet, nothing would have gotten done. So Commerce begrudgingly accepted what NSI had already developed. This ultimately pushed the Internet into a distributed control system. If it had been a fat registry, it would have been a completely different model.

When SAIC acquired Network Solutions in March 1995, there were only about 100,000 domain names in existence. On November 6, 1998, Network Solutions registered its three millionth domain name and, only a year later, that number had doubled to more than six million domain names. NSI built the commercial worldwide domain name business in countries throughout the world and soon became a household name—helping to build the tidal wave of global Internet usage by making it easy to use. The Network Solutions blue-and-yellow globe logo (see Chapter 1, Figure 1-1), which if you look closely is actually two blue gears in a circle on a yellow background, became recognized around the world.

Network Solutions developed a good shared registration system that's still being used today. NSI had a serious potential conflict of interest when the company acted as both domain name registry and registrar, and NSI's management team developed an organizational conflict-of-interest program that was quite effective. NSI put together a full-day, organizational conflict-of-interest course that every employee had to take, from the CEO of the company on down. They really drilled in the fact that employees could not show favorites. And I think they did a good job at that.

Many people found it hard to believe that Network Solutions didn't favor its own in-house registrar, but in my experience it was done well. There was an effective firewall put in place between the registry and registrar functions. In fact, the Network Solutions registrar often felt like NSI's management was being unfair to them because of how strict they were about making sure that they didn't do anything for the Network Solutions registrar that they wouldn't do for their other registrars. Chuck Gomes had to sometimes counsel NSI's customer service team on the registry side that they could and should allow the NSI registrar the same service they would offer any other registrar. And NSI's registry people often bent over backward to serve the non-Network Solutions registrars.

Indeed, I believe that sometimes the Network Solutions people bent over a bit too backward to help their customers. When Gabe Battista took over as NSI's CEO, he found that the company's policies and processes needed much work. This included the handing out of domain names. Says Gabe, "From an operational basis, the whole process of people becoming customers and how they were served was basically not up to snuff, not industrialized. All the simple things that in a mature business, like Telecom was, that had been put in place to increase efficiency, increase customer satisfaction, to ensure that things worked, were all new to this new industry. And so the people that were involved in the industry were learning as they were doing and at the same time trying to drink from a fire hose because the registrations were just flying in."[259]

The SAIC board liked Network Solutions because it delivered significant amounts of profit directly to the bottom line. What the members of the board didn't like was the criticism and unhappiness being generated within the Internet community and the federal government because of NSI's monopoly position as the gatekeepers of the Internet. And we really did take our lumps in the press. For the majority of my tenure at the helm of SAIC, I worked hard to keep the company under the media radar. Despite my ongoing efforts, NSI's monopoly position—and the numerous lawsuits over copyright and other issues—put us squarely on the public stage in a very big way.

I was personally uncomfortable with all the attention—we were used to working on federal government contracts that were often highly classified—and I hoped that it would eventually run its course and fizzle out. It did not, and the political and media spotlight continued to increase in intensity. NSI was regularly assailed in the press for everything from its government-sanctioned monopoly over domain name registration, to the $100 fee it charged to register a domain name, to its censoring of offensive domain names (when NSI blocked the registration of Shitakemushrooms.com because the first four letters of the domain name were considered to be obscene, yet another media storm ensued)…and much, much more. NSI was up to its neck in lawyers, hearings, meetings, government filings for public notice, and all sorts of things that constantly drew focus away from the day-to-day running of the company.

Our offices were flooded with media phone calls and requests for interviews and statements on a variety of events and issues. When we didn't jump into the court of public opinion with our own strong response, we in essence deeded the space to the people who were criticizing us. This turned out to be a mistake, and it is one I regret. The constant pressure of being in the media spotlight wore me down personally; it wore us down as a company too, and it was a huge distraction to our customers.

Ultimately, the SAIC board of directors urged me to agree to sell Network Solutions—primarily because all the negative publicity was beginning to hurt our other businesses. We were looked at as being greedy, and the SAIC culture was definitely not one of greed. We were not a self-serving company. We took tremendous pride in providing solutions for our customers—many of whom worked on projects of vital national importance, which was clearly the case with Network Solutions and its stewardship of the Internet. I reluctantly agreed that selling NSI was the right thing to do, but only if we could get the right price for this valuable asset. Eventually, we were indeed offered the right price, and we decided that the time was right to sell Network Solutions.

Figure 15-2. Photograph of Mike Daniels (right) with Egyptian President Hosni Mubarak (center) and Virginia Governor Jim Gilmore (left), 1999. Reproduced with permission from Michael A. Daniels.

PART SIX

SELLING AT THE PEAK AND LOOKING TO THE NEXT OPPORTUNITY

find the theory of unintended consequences, as popularized by the soci-
ologist Robert Merton, to be of great interest. In short, this theory states
that the outcomes that result from a purposeful action are sometimes not
the outcomes that were originally intended. Further, there are three possible
kinds of results predicted by this theory: a positive unexpected benefit, a negative unexpected detriment, or a perverse effect contrary to what was originally intended.

In the case of SAIC and Network Solutions, we were under increasing
pressure to sell NSI as a result of the spotlight aimed squarely at us by politicians, Internet purists, government regulators, and the press (detailed in the
previous chapter). As a direct result of this pressure—and in response to our
desire to make improvements to NSI's Internet infrastructure, and eventually
to fix a value on the company—we made the decision to sell Network Solutions. We didn't jump fully into the divestment process right away, however.
Instead, we conducted an initial public offering (IPO) in 1997, followed by a
secondary public offering in 1999 and a follow-on offering in 2000. Finally,
soon after we completed the follow-on offering, we reached an agreement
to sell Network Solutions to VeriSign Corporation for the staggering price of
$19.3 billion.

The confluence of two events generated economic value accrued to SAIC
and its employee-owners: urgent customer need combined with the inability
of the National Science Foundation to continue to bear the rapidly increasing costs of domain name registrations. Coupled with SAIC's strong desire to
convert NSI's cost-plus cooperative agreement into a viable business model
driven by a strong and never-ending revenue stream derived from user fees,
the result was the creation of more than $19 billion in value to NSI's owners
and investors when the company was acquired by VeriSign.

While the story of the sale of Network Solutions is a remarkable one, and
it can inform the decisions that business leaders make today, it is just one-half
of the business success equation. The other half of the equation is identifying
new opportunities and then vetting and acting on them. We then arrive full
circle from where we started: the identification of the opportunity that was

Network Solutions by Mike Daniels—an alert technical manager with SAIC—his fight to get our board to agree to the acquisition, and then his personal involvement in turning the company around and helping to make it a prime candidate for acquisition itself just five years later—creating billions of dollars of value in the process.

To achieve long-term success, any serious technology company must be constantly on the lookout for opportunities and then be prepared to act on them. In this final part of the book, we'll explore SAIC's sale of NSI to VeriSign, and the impact of the sale on SAIC. I will also take a look ahead to future opportunities and challenges that every technology entrepreneur and executive should be aware of.

16

SELLING NETWORK SOLUTIONS AND THE AFTERMATH FOR SAIC

Security software maker VeriSign today said it agreed to acquire Net name registrar Network Solutions in an all-stock deal worth about $21 billion.

— CNET News, March 7, 2000[260]

They (NSI) have a pre-eminent position in what they're doing. Now, it gives us an entry point that we can target our enhanced services at. We think the combination will be one of the key infrastructure plays on the Internet.

— Stratton Sclavos, former chairman and CEO, VeriSign[261]

It's a surprise deal but a phenomenal deal. This is the yin and yang coming together for e-business solutions.

— Paul Merenbloom, Analyst, Prudential Securities, Inc.[262]

While the acquisition of Network Solutions, Inc., turned out to be a once-in-a-lifetime opportunity for SAIC and for those of us who worked with the company, it was also a growing headache for our corporation. While I was at the helm of SAIC, my personal preference was to keep publicity to a bare minimum and to generally fly under the radar of the media. As we saw in the previous chapter, this was an impossible task with all the negative press Network Solutions received from its detractors.

At the same time, Network Solutions was a remarkably vibrant business, and it created large amounts of value for SAIC and its employee-owners.

Because Network Solutions was privately held among the company's (and SAIC's) shareholders and not a public company, there was no way to put an accurate value on its worth. As hard as I fought to keep SAIC private and not subject to the short-term demands that would likely be thrust upon it if we went public, I realized that the best path for Network Solutions to take was to prepare it for a public stock offering. In this way the markets would set an accurate value for the company, and we would also generate cash that could be used for a variety of purposes.

In 1999, Michael Lewis wrote a book titled *The New New Thing: A Silicon Valley Story.* The book was the story of serial entrepreneur Jim Clark and his uncanny ability to identify, develop, and take to IPO one great new technology and Internet venture after another. Starting with Silicon Graphics in 1982, Clark went on to found Netscape with Mosaic web browser coinventor Marc Andreessen in 1994 (officially kicking off the dot-com boom of the 1990s), Healtheon in 1998, and MyCFO in 1999. Throughout this period, Clark was a master at identifying the next big technology trend on the horizon, commoditizing it, and taking it to market. In the process, Jim Clark made billions of dollars for himself and his investors. As a direct result of Clark's tremendous success, dot-com businesses took the world by storm, investors flocked to them like pigeons to a bag of birdseed, and any entrepreneur worth his salt snapped up domain names and started as many Internet ventures as possible.

But while it seemed like every company that Jim Clark touched soon turned to IPO gold, Silicon Valley during the mid-to-late 1990s was replete with stories of many other entrepreneurs who also found dot-com success with ideas for Internet-based businesses such as Google, Yahoo!, eBay, Pets.com, Excite, Flooz.com, @Home, eToys.com, Webvan, Kozmo.com, and a seemingly never-ending parade of others. In the preface to *The New New Thing,* Michael Lewis describes the scene:

> In the second part of the 1990s Silicon Valley had the same center-of-the-universe feel to it as Wall Street had in the mid-1980s. There was reason for this: it was the source of a great deal of change. Up until April 4, 1994, Silicon Valley was known as the source of a few high-tech industries, and mainly the computer industry. On April 4, 1994, Netscape was

incorporated. Suddenly—as fast as that—Silicon Valley was the source of changes taking place across the society....The financial success of the people at the heart of this matter was unprecedented. It made 1980s Wall Street seem like the low-stakes poker table. As yet, there is no final reckoning of the wealth the Valley has created. Hundreds of billions of dollars, certainly, perhaps even trillions. In any case, "the greatest legal creation of wealth in the history of the planet," as one local capitalist puts it.[263]

In this red-hot crucible of entrepreneurial creation, dot-com venture after dot-com venture was started up, pitched to potential investors, then sold off as quickly as possible in an initial public offering, enriching the company's founders, early employees, and venture capitalists—and creating a dot-com boom that attracted billions in investor cash. As stock prices for dot-coms shot up—often resulting in market caps far in excess of long-established "old economy" companies such as General Motors, IBM, and Procter & Gamble—investors got caught up in the frenzy, and traditional investment metrics (such as P/E ratio) were ignored. Indeed, the business strategy of the typical dot-com was to capture as much brand awareness and market share as it possibly could and as fast as it possibly could. This, it was understood and assumed by investors, would result in large financial losses for a prolonged period of time. Thus, when it came to profitability, these new-economy dot-com businesses were given a pass in the near term in hopes that the profit would come later.

SAIC had always been a steadily growing, successful, and privately held employee-owned business. We never considered a public offering of our own stock, preferring to keep it firmly in the hands of our employee-owners. (This changed a couple of years after I retired in 2004, when SAIC made an initial public offering of seventy-five million shares of company stock on October 13, 2006.)[264] When we first began to think about taking Network Solutions public, the primary reason was to raise the kind of money necessary to make much-needed improvements in the company's Internet infrastructure and to further fuel NSI's growth.

I have long been personally opposed to taking companies public. From my many years of experience with SAIC, I have found that employee ownership offers considerable benefits to the companies that practice it. Employees are happier and more engaged in their work, providing better products and services to clients and customers. I believe employee-owners are also more entrepreneurial—creating great value for their companies and, ultimately, for themselves. But as much as I was personally opposed to taking a company public, I felt in this particular case that it was best to have Network Solutions publicly owned. I didn't feel that SAIC could continue to take the considerable heat we felt from the Internet community, from the press, and from Congress and the White House if we maintained our ownership of the company. And we thought that if we went into the public marketplace, we would be able to raise the kind of capital we needed to put back into the company because it was growing exponentially and showing no signs of stopping.

We at SAIC knew we had something quite valuable in Network Solutions. The problem was there was no way to assess just how valuable an asset we had since NSI's stock was not publicly traded. So, in part to derive a realistic valuation for the company and in part to test the market, we decided to move forward with a public sale of NSI stock.

In preparation for the IPO, I was determined to learn as much as possible about the technical side of how Network Solutions did what it did. This information would be important for me to know as we spoke with potential underwriters. Mark Kosters recalls one trip I made to NSI's offices to learn more about the workings of the InterNIC. Says Kosters, "One day I was in the office working away and Don [Telage] came by and said, 'Mark, I want you to meet Dr. Beyster.' I was wearing a T-shirt and jeans or something like that, and I said, 'Okay, Don, I'll do that, but don't you think I should go home and change first?' He said, 'No, no—I want you to go exactly how you are now, and I want you to do a presentation on the InterNIC and what it does.' So we sat down— Dr. Beyster, Don, and me—and I explained the InterNIC to Dr. Beyster, how we did our cost recovery, and where our money was coming from. Dr. Beyster asked me questions about the organization, and I remember him taking copious notes as I was talking."[265]

On September 26, 1997, Network Solutions conducted an initial public offering of its stock, with Hambrecht & Quist serving as lead underwriter and J.P. Morgan and PaineWebber assisting. The original plan was for Network Solutions alone to offer up 2.3 million shares of company stock for sale, at a target price in the range of $14 to $16 a share. According to an SEC filing on September 25, 1997, Network Solutions raised the number of shares of stock it offered for sale—to a total of 3.3 million shares—and its corporate parent Science Applications International Corporation also joined in the IPO, offering 500,000 shares of Network Solutions stock that it owned and reducing its ownership interest in NSI to 76 percent. In addition, the target price was raised to a range of $17 to $18 a share.[266]

When trading opened on the morning of September 26, 1997, the first trade was executed at $25 a share, and the share price reached a high of 26-3/4 before settling down at 23-5/16 at market close. This was over $5 more than the high end of the trading range of $18 a share. The IPO raised $50.4 million for Network Solutions and $9 million for SAIC. Perhaps even more important, it established a market cap for NSI. Network Solutions now had a proven value of $382.5 million based on all its outstanding shares of stock.[267]

Considering the fact that Network Solutions was under legal and regulatory assault from many different directions, this was a strong showing. The House Science Basic Research Subcommittee was conducting hearings at the time on the future of the domain name system (a future that presumably could take away NSI's valuable domain name business). In addition, NSI's agreement with the government to register second-level domain names was set to expire in March 1998, and the IPO took place just two and a half months after much of the Internet went dead when an NSI technician loaded a null zone file onto the company's A server—revealing the vulnerability of the Net.[268]

Later, on February 8, 1999, there was a secondary public offering of Network Solutions stock. This time NSI sold 4,580,000 shares of its Class A stock to the public, and SAIC sold 9,000,000 shares of NSI stock, further reducing its ownership interest in Network Solutions to 45 percent.[269] The sale raised $779 million in what was then the largest-ever Internet equity offering. A follow-on offering in early 2000 raised $2.3 billion, again raising the bar of what was at that time the largest Internet offering in history.

On March 10, 2000, the NASDAQ peaked at 5048.52, a high-water mark for the index that was more than double its value twelve months before and which reflected the irrational exuberance of the dot-com boom.[270]

And on March 7, 2000, just three days before the NASDAQ reached its historical peak, e-commerce authentication, payment, and validation provider VeriSign, Inc., of Mountain View, California, issued an SEC Form 425 to announce its intention to acquire a company that was in a unique position in the dot-com boom.

VeriSign's Form 425 announced the pending acquisition, creating a splash in Silicon Valley larger than any dot-com deal had generated up to that point:

Mountain View, CA & Herndon, VA, March 7, 2000—VeriSign, Inc. (Nasdaq: VRSN), the leading provider of Internet trust services, and Network Solutions, Inc. (Nasdaq: NSOL), the world's leading provider of Internet domain name registration and global registry services, today announced the signing of a definitive agreement for VeriSign to acquire Network Solutions in an all-stock purchase transaction. This transaction combines two infrastructure leaders whose trust services support businesses and consumers from the moment they first establish an Internet presence through the entire lifecycle of e-commerce activities.

Under the agreement, VeriSign will issue 2.15 shares of VeriSign common stock for each share of Network Solutions stock as constituted prior to the 2-for-1 split of Network Solutions stock to be completed on March 10, 2000. The transaction, valued at approximately $21 billion based on yesterday's closing price of VeriSign common stock, has been approved by both companies' Boards of Directors and is subject to approval by VeriSign and Network Solutions stockholders. The acquisition is expected to close in the third quarter of 2000, subject to customary conditions, including obtaining necessary regulatory approvals. The resulting company expects to add to its existing employee base to exploit new market opportunities. At closing, Network Solutions will become a subsidiary of VeriSign with Jim Rutt continuing to serve as Network Solutions' CEO, reporting to Stratton Sclavos, president and CEO of VeriSign.[271]

On March 10, 2000—just three days after its announcement—the deal closed, and the owners of Network Solutions sold all the outstanding capital stock of the company in exchange for 72,334,364 shares of VeriSign common stock. VeriSign also assumed options to sell an equivalent total of 8,064,487 shares of VeriSign common stock in exchange for all issued and outstanding Network Solutions options.[272] When the smoke cleared, VeriSign purchased Network Solutions for an estimated $19,277,678,000, of which SAIC's remaining stake was $3.4 billion. The financials of the final deal broke down as follows:

	(in thousands)
Common stock	$72
Fair value of Network Solutions options assumed	$1,426,582
Additional paid-in capital	$17,801,024
Transaction costs	$50,000
Total purchase price	$19,277,678[273]

This made the Network Solutions deal the largest Internet business acquisition up to that time. On the day that VeriSign announced the deal to purchase NSI, the stock price had an immediate impact on both companies. In the case of Network Solutions, the company's shares climbed about 13 percent during the course of the day—from $360.63 at market opening to $407.38 at market close—and hitting a fifty-two-week high of $437.13 during trading. In VeriSign's case, the market was not quite so enthusiastic. In fact, on March 7, 2000—the day VeriSign announced its deal to acquire NSI—VeriSign's shares fell 19 percent, from $247.44 at the beginning of trading for the day to $200 at market close.[274]

Considering the $4.7 million acquisition price, the $3.4 billion we received from the sale of NSI to VeriSign was an excellent return on investment for SAIC's employee shareholders—and one that was not anticipated when the company acquired Network Solutions. And if there was ever an example of fortunate timing, this was it. Within two weeks after the sale to VeriSign was finalized, the air began to leak out of the dot-com bubble, and the Internet technology market began to crash. We had no idea prior to the sale that this would occur, but it did, and in retrospect it was a good thing for SAIC and for our shareholders that we sold NSI when we did.

Today, VeriSign still operates the largest domain name registry in the world, managing more than 50 million digital identities in over 350 languages. As the authoritative directory provider for all .com, .net, .cc, and .tv domain names, VeriSign processes over fifteen billion interactions each day. The company spun off Network Solutions, which is no longer the registry of the Internet. Today NSI functions as a registrar—selling domain names in competition with many other companies—with about 6.65 million domain names under its control, ranking it fourth among the more than one thousand registrars worldwide, after Go Daddy, eNom, and Tucows.

While SAIC was paid handsomely by VeriSign when it acquired the company, VeriSign didn't fare so well when it decided to sell NSI to private equity group Pivotal Private Equity of Phoenix, Arizona, on October 17, 2003. The price? One hundred million dollars for an 85 percent stake, or about $19.5 billion less than it paid for the company just a few years earlier.[275] Before selling Network Solutions to Pivotal Private Equity (since renamed Najafi Companies), VeriSign stripped out NSI's crown jewel (and cash cow), its registry business. According to a VeriSign press release announcing the sale of NSI:

> The Registry business that is the backbone of the global .com and .net domain name infrastructure currently handles over 10 billion interactions per day, remains with VeriSign as a critical component of its business. This Registry business was recently renamed VeriSign Naming and Directory Services and is a core piece of VeriSign's Internet Services Group.[276]

And in 2007, Najafi Companies sold Network Solutions to private equity firm General Atlantic for about $800 million.[277]

As it turned out, a number of the acquisitions that VeriSign made at about the same time as the acquisition of Network Solutions performed poorly, and I believe there are lessons to be learned from this situation. One lesson that stands out is the danger of buying a lot of little companies when you have a highly successful primary business, which takes significant resources away from the main business to finance the acquisitions and assimilate them into your organization. If you make such acquisitions without carefully examining each and every one of them, and if they are in unrelated areas, they can do serious damage to your mainline business. And that's what I believe happened, namely that VeriSign made so many bad decisions under former chairman and CEO Stratton Sclavos that it couldn't sell them off fast enough. And when it did sell the companies, it had to do so at fire-sale prices.

Regardless, VeriSign continues today, as does Network Solutions. In an ironic twist, in 2010 VeriSign decided to sell off its heritage security certificate business. This means that today VeriSign is the true embodiment of the more valuable parts of the Network Solutions business that SAIC owned from 1995 through 2000.[278] Finally, it was announced in August 2011 that Web.com Group, Inc., had reached an agreement to buy Network Solutions, paying $405 million in cash and issuing eighteen million shares of common stock (worth about $155 million) to buy the web domain registrar.[279]

And although SAIC obviously came out very well financially in its sale of NSI to VeriSign, it was not unscathed by the event. The good news was that the sale of Network Solutions to VeriSign was a success for SAIC. The bad news was that the sale of Network Solutions to VeriSign was a success for SAIC.

The story of SAIC's long-term success is perhaps most notable for our "get rich slow" approach to growing the business—one contract and one employee at a time over a period of many years. The company's culture was centered around hiring bright and entrepreneurial people and identifying the needs of our customers. This culture was short-circuited by the acquisition, growth, and subsequent sale of Network Solutions within the span of only five years. Somewhat suddenly, we created an immense amount of profit (roughly equivalent to the entire earnings of the company up to that time) that far exceeded anything else we had ever done before. Our stock price rose rapidly during the time we owned NSI, increasing about five times in five years or so.

This caused a liquidity problem for SAIC—especially as the founding generation of the company's employees began to exit.

Steve Rockwood—who served as an SAIC executive vice president and member of the board of directors—described the financial dilemma SAIC found itself in as follows: "In a closed system like SAIC's, if the ratio of the market cap of the company, which is stock price, to the payroll base of the company, which is your buying power, gets too far out of balance, then you have a problem. And for SAIC, I think the defining moment was Network

Solutions. It pushed our stock price to a very high level. We were all delighted with that, but it didn't add new buyers—no new employees joined the company. So we had a situation where the price of the stock—or market cap of the company—was out of balance with the buying power of the company, which was its payroll. In hindsight, the way to handle the situation we found ourselves in with Network Solutions—the case of hitting a home run when some subsidiary goes out and becomes public or in one way or another generates a whole lot of cash for the company—is pay it to your employees as a dividend. You don't want to run it through your stock system because it will inflate your price relative to your buying power."[280]

While issuing a one-time dividend to stockholders using the funds generated from the sale of NSI might or might not have been a good idea in hindsight, paying a dividend wasn't on the table. During the years I led SAIC, the company never once paid out dividends to stockholders.

Clearly, things were different at SAIC after our experience with Network Solutions—good in some ways, and not so good in others. Because the financial success of NSI was for the most part disconnected from SAIC's core, and because it all happened so quickly, we created an environment that defocused the company in subtle ways and very well may have led to the dilution of our unique culture. This in turn may have at least in part set the stage for the October 2006 initial public offering that overnight transformed SAIC from a company entirely owned by its employees into a public company largely beholden to Wall Street analysts, quarterly earnings reports, and the whims of public opinion.

17

ONGOING CHALLENGES AND OPPORTUNITIES

The future of business is not the stuff of science fiction; it's the extrapolation of shifts that are beginning to take place right now before your eyes. And while there's nothing you can do to prevent it, there are things you can do to prepare to thrive in it that will deliver real business value today, and give you a competitive advantage as changes take place.

— Steve Mills, senior vice president and group executive,

IBM Software Group[281]

Although much of the story of the birth and tremendous growth of the Internet is now history, the story is not over yet. The Internet continues to offer both opportunities and challenges. This final chapter focuses on the opportunities and challenges associated with the domain name system (Section 1), and with the web and the Internet (Section 2).

Section 1: Opportunities and Challenges with the Domain Name System

The evolution of the domain name system continues today, more than thirty years after it was originally devised. In this section, we'll explore the new generic top-level domains, as well as major DNS policy concerns that need to be addressed to ensure that the Internet continues to work smoothly.

ICANN's New Generic Top-Level Domains

American businesses have long made the .com top-level domain their pre-ferred place to host websites and to conduct online commerce. However, this hasn't stopped the Internet powers that be from trying to induce them to move into new top-level domain names. As we noted in Chapter 15, as early as 1996 Jon Postel floated the idea of creating a new set of top-level domain names—at least in part to break NSI's monopoly over domain name registra-tion. Since that time, ICANN has rolled out two new sets of gTLDs. In 2000, ICANN introduced .aero, .biz, .coop, .info, .museum, .name, and .pro, and in 2004, the organization introduced .asia, .cat, .jobs, .mobi, .post, .tel, .xxx, and .travel.[282] For the most part, these new gTLDs have had only limited success—certainly nothing on the order of .com—and because of political opposition, implementation of the .xxx gTLD was shelved until 2011.

However, there is something new in the world of generic top-level domains, and it is creating great opportunity along with great controversy.

In a June 20, 2011, vote, ICANN's board decided to authorize the launch of a new gTLD program—a program that would dramatically expand the number of gTLDs available. In the works since 2005, ICANN's new gTLD program is designed to allow individuals and organizations—from entrepreneurs, to businesses, to governments, to entire communi-ties—anywhere in the world to introduce and operate their own unique top-level domains.

So, for example, Ford Motor Company might decide to apply for the gTLD .ford, which the company could then use in lieu of (or in addition to) its long-established ford.com URL. And if Ford so desired, it could act as the registrar of the .ford domain, selling (or giving away) .ford-based URLs as it pleases. For example, Ford could give out unique URLs to its dealers, such as Mossy.ford for the San Diego-based Mossy Ford dealership, or Koons.ford for the Falls Church, Virginia-based Koons Ford dealership.

Similarly, an Internet entrepreneur or company could buy a potentially pop-ular gTLD and then act as a registrar for the domain, setting a price for its URLs and operating the registration process. For example, an individual could buy the new gTLD .scuba, and then sell unique domain names based on the .scuba top-level domain to scuba diving schools, equipment retailers, travel agencies, organizations, and individuals.

Of course, it seems that whenever ICANN proposes to make a change to the way the Internet works, controversy is never far behind. That is indeed the case with the new gTLDs. A number of organizations and members of the Internet community have spoken out against the plan. In one notable example, ICANN has attracted the ire of the Association of National Advertisers (ANA) which believes the new program will be bad for business.

Says ANA president and CEO Bob Liodice, "By introducing confusion into the marketplace and increasing the likelihood of cybersquatting and other malicious conduct, the ICANN top-level domain program diminishes the power of trademarks to serve as strong, accurate, and reliable symbols of source and quality in the marketplace. Brand confusion, dilution, and other abuses also pose risks of cyber predator harms, consumer privacy violations, identity theft, and cyber security breaches."[283]

In addition, the ANA and other trade organizations and businesses are concerned that they will be forced to snap up numerous gTLDs to protect their established and future brands and trademarks. For example, continuing with our example of the Ford Motor Company, buying the .ford top-level domain is just the start. To fully cover its many different brands, Ford would also need to attempt to lock down a variety of other gTLDs, including .lincoln, .mercury, .mustang, .fiesta, .taurus, .focus, .ranger, and on and on. The legal wrangling over some of these gTLDs as individuals and organizations fight to protect or assert their brands and trademarks will likely be epic. And with an "evaluation" fee of $185,000 for each new gTLD, plus an ongoing annual fee of $25,000, many individuals, small businesses, and nonprofits will be out of luck when it comes to obtaining their own new generic top-level domains.

But according to some observers, these fees are just the tip of the iceberg. Says Scott DeFife, executive vice president of policy and government affairs for the National Restaurant Association, "Restaurants of all sizes will be forced to apply for new domains to protect their brands and trademarks. Costs include a $185,000 application fee for each new top-level domain. Restaurants and other companies also likely would be forced to register numerous second-level domains—the words to the left of the 'dot' in domain names—within the new top-level domains. Costs would be driven higher by legal, marketing, and other costs. Some businesses have put the cost of registering a single top-level domain at $2 million or more over the initial ten-year contract as companies submit applications, watch and defend their domains, monitor for infringement, and litigate to block abuse. Costs could run higher if businesses are forced to buy their own Internet names in auctions."[284]

Responding to the concerns voiced by ANA and others, the US Senate Committee on Commerce, Science, and Transportation called a hearing on December 8, 2011, to discuss ICANN's plans to examine the merits and implications of ICANN's new program and the organization's continuing efforts to address concerns raised by the Internet community.[285] In the meantime, ICANN is moving forward with its plans and it expects to have the first new gTLDs in operation by 2013.[286]

Major DNS Policy Concerns Going Forward

The draft final report of the WHOIS Policy Review Team, published in December 2011, presents a critical assessment of the significant problems ICANN faces going forward in policy codification and in maintenance and enhancements of the Domain Name System (DNS). The sixteen-person review team comprised the ICANN CEO, the chair of the Governmental Advisory Committee (GAC), and other international DNS stakeholders and experts. A brief recap of the report follows.

WHOIS is chartered by ICANN to maintain and provide public access to the contact information submitted with domain name registration. WHOIS was initiated in the early '80s, before the commercial Internet began. The three components of WHOIS are:

1. WHOIS Data: The information provided by DNS registrants when registering a domain name.
2. WHOIS Protocol: The standard communications elements used for queries and responses for public access to the data.
3. WHOIS Service: Methods offered by registries and registrars for public access to the data.

The ninety-two-page WHOIS report finds that there is ongoing emotional debate over serious problems with major aspects of WHOIS in important areas such as policy, data accuracy, privacy, proxies, and cost. The report concludes that "the current system is broken and needs to be repaired."

The WHOIS Policy Review Team presented eighteen recommendations for improving the WHOIS service. The first four pertain to policy.

Policy

The report concludes that ICANN's policy is poorly defined and recommends that the policy be codified in one place, and that the WHOIS process for maintaining data and providing public access should be made an ICANN strategic priority.

Accuracy

The next five recommendations focus on improving data accuracy. In the period 2009-2010, ICANN sponsored a data accuracy study, which was conducted by the National Opinion Research Council (NORC). The NORC WHOIS study found that only 23 percent of WHOIS records were fully accurate, and that more than 20 percent were completely inaccurate. The conclusion of the study stated that this low level of accuracy is unacceptable and that it decreases consumer trust in WHOIS data and in the industry itself.

The Review Team recommended that ICANN:

- Undertake appropriate measures to reduce the number of unreachable WHOIS registrations by 50 percent within twelve months, and another 50 percent in the next twelve months.
- Publish an annual accuracy report indicating the degree of measured improvements achieved.
- Provide an annual status report on progress toward the goals set forth by the WHOIS review team.
- Ensure that there is a clear, unambiguous, and enforceable chain of contractual agreements with registries, registrars, and registrants requiring the provision and maintenance of accurate WHOIS data.

- Ensure that the requirements for accurate WHOIS data are widely and proactively communicated to current and prospective registrants.

Privacy and Proxy Services

The next seven recommendations focus on privacy and proxy services that limit public access to information about domain registrants. A privacy service provides the registrant's name but only a subset of other WHOIS information, in order to limit public access to full identity. A proxy service is where an agent serves as the registrant, acting on behalf of another, thus disguising the actual registrant's true identity.

The Review Team observed that the use of these two services is widespread. A 2010 study found their use in 15 to 20 percent of the registrant records, or 20 million gTLD DNS registrations. These services, offered by a wide range of service providers, including some registrars, have been subject to long-standing and ongoing vigorous debate by the various stakeholders arguing over the balance between privacy and legitimate public access requirements in an unregulated DNS environment. Stakeholders with legitimate concerns include global law enforcement agencies, entities engaged in international privacy and intellectual property laws and regulations, major businesses, and various advocacy organizations.

The report provides numerous comments from the various stakeholders. For example, a law enforcement agency stated, "Proxy services play into the hands of organized crime, they hide all their business behind them and this is a huge issue, not only for law enforcement, but for the wider Internet community as a whole."

A large hotel group stated, "Privacy services have frequently frustrated our ability to protect our hotel brands online, which, unfortunately, often leads to confusion and other problems among consumers."

The Coalition for Online Accountability stated, "Until ICANN is able to bring some semblance of order, predictability, and accountability to the current 'Wild West' scenario of proxy registrations, it will be impossible to make significant progress toward improving the accuracy of WHOIS data, so that the service can better fulfill its critical function to internet users and society as a whole."

Conversely, numerous comments were made supporting the need for privacy and proxy services and proposing that these services be accredited and regulated. The WHOIS Policy Review Team presented seven recommendations to ICANN regarding privacy and proxy services:

- Develop and manage a system of clear, consistent, and enforceable requirements for all privacy services consistent with national laws.
- Develop a graduated and enforceable series of penalties for privacy service providers who violate the requirements with a clear path to de-accreditation for repeat, serial, or otherwise serious breaches.
- Facilitate the review of existing practices by reaching out to proxy providers to create a discussion that sets out current processes followed by proxy service providers.
- Registrars should be required to disclose their relationship with any affiliated retail proxy service provider to ICANN.
- Develop and manage a set of voluntary best practice guidelines for appropriate proxy services consistent with national laws. These voluntary guidelines should strike an appropriate balance between stakeholders with competing but legitimate interests; at a minimum this would include privacy, law enforcement, and the industry around law enforcement.
- Encourage and incentivize registrars to interact with the retail service providers that adopt the best practices.
- Include an affirmative statement that clarifies that a proxy means a relationship in which the registrant is acting on behalf of another. The WHOIS data is that of the agent, and the agent alone obtains all rights and assumes all responsibility for the domain name and its manner of use.

Data Access-Common Interface

The WHOIS Policy Review Team's final four recommendations focus on making improvements for access to WHOIS data by means of a multilingual website, final data model for international services and local language requirement, and metrics for accuracy and availability:

- Set up a dedicated multilingual interface website to provide thick WHOIS data to improve access to the WHOIS data of .com and .NET gTLDs—the only remaining thin registries. An alternative for public comments is also provided. To make WHOIS data more accessible for consumers, ICANN should set up a dedicated, multilingual interface website to allow "unrestricted and public access to accurate and complete WHOIS information." Such interface should provide thick WHOIS data for all gTLD domain names.
- Task a working group within six months of publication to finalize encoding, modifications to data model, and international services, to give global access to gather, store, and make available internationalized registration data.
- Reflect and incorporate the final data model and services in registrar and registry agreements within six months of adoption of the working group's recommendations by the ICANN board.
- Finalize requirements for registration data accuracy and availability in local languages along with the efforts on internationalization of registration data, define metrics to measure accuracy and availability of data in local languages and (if needed) corresponding data in ASCI, and explicitly define compliance methods and targets accordingly.

The Challenges Continue

The WHOIS report shows clearly that major challenges in domain name policy and administration continue—particularly in achieving better data accuracy, balancing legitimate privacy and public access legal and ethical requirements, providing multilingual access, and avoiding fraud and misuse via proxy services.

Section 2: Opportunities and Challenges with the Web and Internet

The continued growth of the Internet is upon us. As the example of Network Solutions demonstrates, technological change during periods of exponential market growth is extremely rapid with Internet-based businesses and influenced by many nontechnological factors (e.g., socio-politics, standards and governance, values). I see the rate of change only accelerating and becoming more intertwined with social, economic, and political factors.

Because NSI's employees were at the forefront of the emerging Internet—and they had the smarts to quickly adapt to the rapidly changing technology—Network Solutions became an early leader in the fast-growing Internet industry. The company's sharp focus on responding to the growing needs of the company's quickly expanding customer base and on strengthening and growing the infrastructure of the Internet systems for which it was responsible also helped to cement its position as an industry leader.

But it would be a mistake to underestimate the other, nontechnical smarts it took to turn Network Solutions into a viable business, unlimited by the shortsighted focus of people who couldn't see the big picture. To thrive over the long term, technology companies must combine technical and business smarts with an inspiring (and correct) vision of the future. When we acquired NSI, the company was long on technical skills, but short on business skills and vision. At SAIC, we had a deep bench of business expertise from which to draw, and we brought it to bear at NSI, quickly turning around the company's fortunes.

As technological innovation continues to accelerate in the future, change will surely continue to be difficult to predict for technology entrepreneurs. As a result, the ability to anticipate and then respond to a range of changes before they arrive will become an increasingly important characteristic of any successful technology organization.

The best first step to becoming an organization that can readily adapt to dynamic change is to hire talented, technically savvy men and women who are able to ensure a solid technological foundation for the business, and who can convert technological change into organizational action. These people are intimately familiar with advances in technology, they don't cut corners or shy away from tough technical problems, and they provide you with an early warning system as change approaches your industry from over the horizon. In fact, they help you create change by redefining problems such that new solutions are available. Because these solutions often reach deeply into our social fabric, such as concerns over the privacy of users of online applications and websites, shaping and commercializing these solutions will increasingly take a multi-disciplinary team that brings the necessary savvy in the social and political sciences, which is a part of any major Internet-based business today.

I believe you should seek out technology areas that are undergoing rapid change. This position is naturally one of great leverage—both toward significant success and significant failure. Network Solutions was dead center in the maelstrom of change surrounding the growth and commercialization of the Internet. Today, many technology areas are experiencing similar growth and change. My coauthor Mike Daniels and I believe that the following great technology waves are shaping—and will continue to shape in the future—the growth and evolution of the Internet industry:

- Social. This includes the proliferation of social media such as Facebook, Twitter, and YouTube, which have lately taken the Internet by storm and have created a vast amount of value in a short amount of time.
- Mobile. More than 85 percent of mobile phone handsets today are web-enabled, drawing a huge wave of users to the Internet— many of whom don't access the Net through any other technology platform.
- Local. Today there is a strong focus on local—local deals, local places to eat, local stores and businesses, local directories, local maps and directions, and much, much more. People are increasingly using the Net to access local information, and this is driving traffic ever upward for these sites.
- Cloud. Cloud computing is the next big thing when it comes to the confluence of individual computer users and the Internet. In cloud computing, instead of storing data files, photographs, music, and even applications such as word processing or spreadsheet programs on local computers, these items are stored on remote servers maintained by companies such as Google or Microsoft.

Each of these technology waves has demonstrated the ability to scale from small to large—with no immediate end in sight. And while the great technology wave that was the domain name system has flowed and ebbed, it is still undergoing significant changes that have the potential to dramatically impact the ways in which we interact with the Internet.

For example, the Internet enables a computer in one location to communicate with any other computer on the network, whether it's across the hall or on the other side of the globe. However, to accomplish this feat, the Internet must know the location—that is, the specific address—of both source and destination nodes on the network. The Internet Protocol portion of the Internet Protocol Suite provides these addresses, albeit in a way that is not easy for people to recognize or remember. Why? Because it does so in the form of a 32-bit binary number in the case of the original system—Internet Protocol Version 4 (IPv4)—and a 128-bit binary number in the case of the newer Internet Protocol Version 6 (IPv6) system.

However, because the IPv4 system is limited to a total of 4.3×10^9 (approximately 4.3 billion) number combinations, and large blocks of IPv4 are reserved and unavailable for public allocation, it is predicted that the system will soon run out of IP addresses. As a result, there has been mounting pressure from a variety of quarters to convert the present system over to IPv6, which can accommodate 3.4×10^{38} number combinations.[287] At the date of this writing, the issue remains unresolved. (Note: A proposed replacement for the domain name system is the Handle System, developed by the Corporation for National Research Initiatives [www.handle.net], which does not rely on the domain name system to function.)

Just as the application of the domain name system was just one of many possible choices for connecting Internet users with desired destinations, there are many choices to be made as the Net continues to mature. As I look ahead to the future of the Internet, I know that there are many changes that we cannot today anticipate with any degree of accuracy, and that these changes will rapidly continue to come about. Three specific areas within the Internet industry have long-term potential to experience strong, ongoing growth that I would like to briefly address: security, web applications, and ICANN's new generic top-level domains.

Security

Unfortunately, the very thing that makes the Internet such an important global communications medium—its openness—is the Achilles' heel that makes it especially vulnerable to attack. Today, Internet security is a fast-growing concern, and for good reason: an unlimited supply of hackers is trying to push their way into American companies and government installations.

The threat to the Internet comes in many shapes and sizes, and as long as there are smart and motivated hackers in the world, I believe no computer system that has a portal to the Internet will ever be safe from infiltration and potential manipulation. According to Bill Roper, former CEO of VeriSign, Internet cyberattacks are widespread and remarkably frequent. Says Roper, "VeriSign's network—just the segment of the Internet that VeriSign runs— sustains more than two million attacks a day. It's the most attacked network on the face of the globe. That's a huge scale of attacks, and they're growing. They are attacking the network itself, attacking websites, attacking companies, attacking governments, and attacking anything that is visible. They do this in some cases for economic gain, in some cases for political embarrassment, and in some cases just because it's there."[288]

Every day, unknown foreign entities attempt to penetrate computer networks within NASA, the Departments of Commerce, Homeland Security, and other federal government agencies. According to Deputy Secretary Lynn, more than one hundred foreign intelligence organizations are currently attempting to hack into secure US military computer networks—sometimes with success.[289] Even White House computer networks have been the targets of foreign governments and cybercriminals.

The government is not the only target of hackers and cybercriminals. Businesses are also routinely the subject of cyberattacks and intrusions. In 2008, flash drives were used to hack into the computer network of home mortgage lender Countrywide Financial Corporation. Customer account and personal information—including credit card and Social Security numbers—were stolen by the hacker and then sold to a third party. This person then resold the information to Countrywide's competitors and others. The loss to Countrywide was pegged at more than $80 million.

The threats will continue to increase at the governmental and commercial levels and probably spread more widely across the globe. But the threats won't just spread geographically—they will spread technologically. As the

Internet continues to proliferate into new platforms, cybercriminals won't be far behind. If you connect to the Internet, you cannot keep people out of your computer. There's no way to guarantee perfect security for your own computers when they are connected to the public Net—even when you have invoked all the security systems present on your computer.

This fact has immediate and widespread implications for individual computer users and for businesses, nonprofits, and governments—and for the businesses that decide to tackle the security challenge by providing innovative new products and services. Now that anyone with a computer, an Internet connection, and enough technical smarts can breach the security of most any computer or computer system on the Net, most every company, nonprofit, and government entity is exposed to the threat of having its security breached. It's an incredibly inexpensive proposition for a single individual to mount an attack on another computer.

In the case of individuals and smaller organizations, effective security solutions are generally focused on relatively inexpensive or even free virus scanners that run on individual computers, spam filters that block infected messages before they can reach their intended recipients, and "smart" web browsers that catalog and avoid sites with malicious code and other security threats.

Larger organizations and governments often draw the attention of highly motivated professional (and state-sponsored) hackers and cybercriminals, and security solutions are significantly more complex—and expensive—than those for smaller organizations. Effective security solutions in cases such as these require hiring (or contracting with) cybersecurity professionals who develop extensive plans and strategies—and deploy hardware, software, and web-based solutions—to defend the organizations against attack.

Defending against such cyberattacks—which can come from anyplace anytime—can be an expensive proposition, especially for larger organizations, which must defend themselves on multiple fronts. This makes the cybersecurity industry one that will continue to experience high rates of growth for the foreseeable future, and the opportunities to build and grow successful businesses in this area will be significant.

Web Applications

Throughout the history of modern computing, many smart men and women have dedicated themselves to the search for software applications that are so desirable to users that they seemingly can't live without them. The VisiCalc spreadsheet for the Apple II computer—followed by Lotus 1-2-3 for the IBM PC—was an extremely successful business application that drove sales of personal computer hardware.[290] Electronic mail was the killer app that first built public interest in the Internet—and it remains one of the most popular to this day.

The kinds of applications that can be offered to potential users—and their ultimate adoption—are in large part driven by fast-changing computer and Internet technology and in the ability of the infrastructure to support it. Faster computers coupled with increased Internet bandwidth have enabled a rich online multimedia environment to emerge and take hold of the public consciousness. Full-featured web applications—most recently expressed as social media websites such as YouTube, Facebook, and Twitter that support written messages, photos, video, and more—are the result.

I believe that five web applications in particular merit attention—both because they provide effective models of how to leverage Internet technology to create tremendous value, and because I expect them to be around many years from now.

- **Facebook.** This social media application—launched by Harvard dropout Mark Zuckerberg and three of his college classmates in 2004—has in a very short time taken the world by storm. The basic idea of Facebook is to provide users with a place where they can create personal profiles, add other users as friends, and exchange messages. In addition to that basic functionality, users may join common-interest user groups, organized by school, company, city, or a wide variety of other characteristics. At the time of this writing, Facebook has more than 800 million active users[291] and is on a trajectory to surpass one billion active users in the not-too-distant future.

- **Google.** In the mid-1990s, there was an active ecosystem of search engines (a web application that allows users to perform Internet information searches) available for Internet users to use to find the information they sought—everything from Yahoo! to HotBot to Lycos to AltaVista. However, when Stanford PhD students Larry Page and Sergey Brin brought their new search engine Google online in 1997, it quickly left the competition behind. The Google website itself was simple and easy to use, and it consistently provided users with better results more quickly than other search-engine applications. Today, Google processes more than one billion searches each day, and google.com is consistently ranked the number one most-visited website in the world, although Facebook is nipping at its heels.

- **Skype.** Founded in 2003 by Dane Janus Friis and Swede Niklas Zennström, Skype is a peer-to-peer, voice-over-IP (VoIP) application that allows users to talk to one another directly via the Internet—bypassing long-established wired telecommunications networks. And because the basic functionality of Skype (now including video and conference calls) is available to users at no charge, it has become a popular web application with more than 660 million users worldwide—many of whom are business users. The company was acquired by eBay in 2005 and subsequently sold off to a private equity firm in 2010 and acquired by Microsoft in 2011.[292]

- **Twitter.** While the idea of web logging—blogging—has been popular for at least the last decade or so, it took the Twitter web application (launched by Twitter creator Jack Dorsey in 2006) to popularize the idea of microblogging. With Twitter, users can broadcast messages of up to 140 characters (known as tweets) to anyone who has signed up to follow them, and messages are also publicly available for viewing on individual user pages located on the twitter.com website. The content of these tweets varies considerably—from status updates, to links to web information of interest, to messages to other Twitter users, and much more. There were more than 300 million Twitter users as of the date of this writing, broadcasting more than 250 million messages each day.[293]

- **YouTube.** Before the creation of YouTube in 2005 by three former PayPal employees, there was no simple way to upload, view, and share videos on the web. Today, however, YouTube has enabled an explosion of user-generated video content, including film and television clips, music videos, and user-created original videos and video logs (vlogs). The company was acquired by Google in 2006. Many large media companies, including CBS, BBC, VEVO, Hulu, and others have jumped onto the YouTube bandwagon, offering programming via YouTube's partnership program. More than three billion videos are viewed on YouTube each day, and it is the second most popular search engine in the world.[294]

While Google—a search application—is currently the most popular web application, social media is clearly where much value is being created today. In broad terms, social media can be defined as Internet-based applications that allow computer users to easily communicate with one another and engage in interactive dialogs. Social media has shown strong growth over the past several years, as improvements in Internet infrastructure and increased public awareness has increased its market penetration. According to recent statistics,

- Facebook has 665 million active daily users, and 1.1 billion active monthly users.
- 21 percent of the world's Internet population uses Twitter at least once a month.
- YouTube has 1 billion unique users who visit every month, and these visitors watch more than 6 billion hours of video.
- Two new users join LinkedIn every second.[295]
- Google + was the fastest social network to reach ten million users, at sixteen days (Twitter took 780 days and Facebook 852 days).[296]

While traditional media can occasionally outshine social media for brief moments, it cannot do so on a sustained basis. For example, one billion people around the world tuned into the Opening Ceremonies of the 2012 Olympic Games in London—about 15 percent of the world's population.[297] Traditional media do not have the ongoing daily reach that social media does. The newspaper with the largest circulation in the world only draws about fourteen million subscribers,[298] and television and radio networks measure their active viewers and listeners at any given time in the single-digit millions.

Figure 17-1. In 60 Seconds. Reproduced with permission from Shanghai Web Designers.

As technology continues to move forward, and as an increasing number of people are able to access social media through a variety of different platforms, including their phones, computers, and even their DVD players, we can expect the current forms of social media to increase in reach and popularity, while new forms emerge. This will provide ample future opportunities for technology companies to innovate entirely new social media applications, platforms, and worlds. It is up to business leaders and investors to carefully consider which will offer the best opportunities and returns on investment and to pursue them vigorously.

For those businesspeople who hope to find and develop new opportunities in this fast-growing market, six major categories of social media as defined by Professors Andreas Kaplan and Michael Haenlein of ESCP Europe have emerged. These categories include:

- Collaborative projects (Wikipedia, Digg, WordPress)
- Blogs and microblogs (Blogger, Twitter, Foursquare)
- Content communities (YouTube, Flickr, Yelp)
- Social networking sites (Facebook, MySpace, LinkedIn)
- Virtual game worlds (World of Warcraft, EverQuest)
- Virtual communities (Second Life)

Each of these different categories serves different kinds of customers, who in turn have uniquely different needs and wants. So what exactly has driven the unprecedented growth of Facebook? Perhaps it's the fact that the website has tapped into a deep human need: the need to be liked by others. It's no small coincidence that the Facebook "Like" button has broken out of the website's online boundaries and into the popular consciousness. This has attracted the attention of large companies, always on the alert for new ways to reach potential customers.

The Ford Motor Company recently sponsored an online campaign for its 2012 Focus automobile. The focus of the campaign? A free Facebook page hosted by an orange-colored puppet by the name of Doug. By including online conversations and videos starring Doug in this Facebook page, Ford managed to attract more than 43,000 "Likes" from fans of the car. While Ford spent more than $95 million to advertise the new Focus across all the normal media channels, it spent just a small fraction of that total on its Facebook campaign—deriving results far in excess of what it spent.

Facebook is moving quickly to attract potential advertisers and monetize its corner of the web, primarily through the rollout of new applications on its site. For example, Facebook has introduced an application that enables users to connect their accounts with the Spotify (www.spotify.com) online music service. Once connected to Spotify, Facebook displays the name of the artist and the name of the song that a user is listening to. However, it goes one step further: Facebook leverages that information into its advertising feeds. If a Facebook user listens to a song by Elton John on Spotify, then Facebook can feed an ad promoting an upcoming Elton John CD release or concert tour to the user. According to Internet marketing research firm eMarketer, Facebook has taken the lead in US online display advertising, with projected revenues of about $2 billion in 2011.[299]

This is the value proposition that social media and the vast universe of rapidly proliferating web applications offer businesses that choose to make them a key part of their advertising campaigns: lots of visibility with thousands or perhaps even millions of potential customers, for not a lot of money. Even with the significant problems surrounding the Facebook IPO in 2012, we believe these types of social media companies will continue to attract billions of users globally and their managements will find ways to monetize these massive user bases.

History Informs the Future

Soon after we acquired Network Solutions, it became clear that the existing cost-plus contract with the National Science Foundation to administer the DNS was not structured in a way that would allow the Internet to scale, given the exponential cost increases that would be directly transferred to the federal government. We pushed for the establishment of a domain name registration fee—an idea that was vociferously fought by many in the Internet community. The National Science Foundation agreed with our position, however, realizing that its own internal budget could not support the continued free distribution of domain names.

As a result, we established a fee of $50 a year for each domain name. This fee—established in partnership with a key federal government agency—enabled us to develop with Network Solutions a highly scalable and profitable business model that clearly demonstrated to the world that there was great value in this nascent industry.

The climate in Washington has changed significantly in recent years. For as long as I can remember, American business has been, for the most part, a trusted partner of the US government. The government—and, indeed, our nation—has relied on businesses to lead the way in many different areas, including economic growth, innovation, and the kind of productivity and efficiency that make our nation the envy of the world. Unfortunately, the public's trust in the private sector has eroded considerably, and this has ramifications for the future of technology entrepreneurs.

When the National Science Foundation allowed the Internet to move into commercial play and not be smothered by government management and regulation, this became a key decision that enabled a "million flowers to bloom" all around the globe. Thousands of companies have been built solely around the Internet, billions of dollars of wealth have been created, and thousands of new, innovative jobs have been born. I believe that this would have never come about without the government's fundamental decision to entrust the Internet to the private sector. Network Solutions partnered closely with the United States government to help ensure that the commercialization of the Internet we take for granted today moved forward. It remains to this day an excellent example of what government and business can accomplish when they work together.

The government needs to continue to selectively fund and support innovation through the private sector. There will be failures—sometimes spectacularly embarrassing failures—but that shouldn't stop us from trying to create successful public-private partnerships. We won't be able to compete effectively on the global stage if we can't collaborate. This will require that private industry pay its fair share in developing and nurturing new technologies, as it has in the past.

America—with the cooperation of government and the private sector—can and must put its focus on innovation, economic growth, and the creation of quality jobs in this country. Political slogans and rhetoric do not create jobs, and they do not build global industries. Smart investment of many types of capital (e.g., loans, research grants, subsidies, tax credits) by the US government in advanced science and technology—taken by the private sector and then built into successful companies and industries, sometimes over a period of decades—is the place where massive wealth can be created. Economic growth and quality jobs are created by a vibrant, dynamic capitalistic system. It seems that lost in the current debate is any real discussion of what truly works to create economic growth, quality jobs, and wealth. The Network Solutions story provides solid evidence that investment in a business and its people (in this case, by SAIC) and in science and technology (by the public sector, in this case, NSF) can lead to significant economic growth, quality jobs, and wealth creation for investors and employees.

During the early 1990s, the Internet reached a tipping point in its development. While there was much pressure by entrenched interests to maintain its status as a semi-private communication network between universities and government agencies, a few visionaries saw that the Internet had the potential to be much more. The wonders of today's Internet are a direct result of the vision that these men and women laid out. I am certain that tomorrow's Internet will similarly be shaped by today's visionaries. We have just scratched the surface of the potential that this still-young technology holds for the future. And it is indeed a bright future.

The story of the Internet has been told many times by many different people, and my goal in writing this book was not to repeat the previous efforts of others. What is often left out of stories about the origins and growth of the Internet industry, however, is the importance of a small technology company by the name of Network Solutions, Inc., to the evolution of the Internet and its transition from glorified electronic mail system into the essential and pervasive global business and communications powerhouse that it is today. It wasn't just that the obscure area of domain name registration was technically challenging, but NSI also had to deal with difficult sociopolitical factors that one might think were only having an impact on telecom or application companies. After all, what could be controversial about a URL?

Looking back, I had no idea that so many companies would adopt the Internet, and it amazes me even today that these companies continue to proliferate. I did realize that companies needed to become network-centric and to leverage the many benefits in cost, advanced technologies, and other advantages of the Internet extensively for all of their business operations. Regardless, I am truly amazed that the Internet turned out to be the great, game-changing success that it has been. And, of course, the sale of Network Solutions to VeriSign in 2000 was a game changer for us. The proceeds that we derived from the sale provided significant resources to feed the growth of SAIC, all the way through 2003.

SAIC was a loose federation of businesses—what we called our planetary organizational model—held together by tight financial controls. As a subsidiary of SAIC, Network Solutions was a part of this loose federation. While NSI's employee culture was different from ours—more fast-paced dot-com startup than methodical veteran government contractor—we gave it the same autonomy as we did our other subsidiaries. This approach to business served us well—allowing division managers to respond to new business opportunities quickly and without having to constantly ask my permission to proceed before they acted. It also enabled a highly entrepreneurial culture where employees were encouraged to find new business opportunities and then rewarded with ownership in the company when they were successful in converting those opportunities into contracts.

Our experience with Network Solutions reinforced the power of following the basic values by which we ran SAIC: to encourage entrepreneurial spirit while unleashing real value, to give employees unprecedented levels of authority to run their own business areas, and to give those employees the accountability and responsibility to own the company they are growing. It also proved that the best business opportunities often bubble up from below—championed by passionate employees and not as a result of top-down strategic-planning exercise. Individuals can and do make a difference.

Where there are customers, there will always be needs—some of them urgent. For the entrepreneurs and business leaders who are able to discern these needs and then respond, either with entirely new products and services or with offerings that are better, less expensive, and more profitable, I am convinced that the rewards will be great.

There are, of course, no guarantees—in life or in business. For every great new product or service introduced into the marketplace, there are many failures. But this is the nature of business, and of entrepreneurship. We must pick ourselves up, learn from our mistakes, try again, and never give up. I for one will always stand on the side of opportunity, and I encourage you to personally do everything in your power to pursue it.

APPENDIX
A BRIEF INTERNET PRIMER
(FROM THE BEGINNING TO DARPA)

Just as the telegraph and telephone ushered in a revolution in communication in the nineteenth and twentieth centuries, so, too, has the Internet created its own revolution in communication. However, the Internet is having an impact far beyond the previous communications technologies, enabling the development of new, previously unknown global social networks—some for buying and selling goods and services (eBay and Amazon, for example), some for networking with business and professional colleagues (LinkedIn and XING, for example), and others just for connecting with friends and relatives (Facebook, Twitter, and MySpace, for example).

While there is plenty of technology behind the massive interworking of the global Internet, three innovations in particular led to its current form and function. These innovations are packet switching, TCP/IP, and the HTTP.

Packet switching is a technology first invented by Paul Baran of the Rand Corporation. Packet switching is a set of protocols that break messages into small packets before they are sent. The actual term *packet* was invented by Donald W. Davies of the National Physical Laboratory in Teddington, United Kingdom. These packets may follow different routes to their ultimate destination (where they are put back together to re-create the original message), allowing them to circumvent damaged or nonfunctioning communications links or nodes. Leonard Kleinrock's queuing theory of message-switched networks played an important role in showing the efficiencies that could be derived from packet switching.

TCP/IP (acronyms that stand for Transmission Control Protocol and Internet Protocol) is the basic set of rules that govern transmission of data across the Internet. TCP/IP is a two-layer system, with TCP handling the higher-level task of breaking up and reassembling messages into packets that are transmitted across the Internet, and IP handling the lower-level task of addressing packets so they get to the right destination. Vint Cerf and Robert Kahn codesigned the second-generation DOD Transmission Control Protocol suite.

Hypertext transfer protocol (HTTP, developed by Tim Berners-Lee and Robert Cailliau) is the set of rules that govern the transmission of text, graphics, audio, video, and other files over the World Wide Web. The development of HTTP—along with a reliable web browser—led directly to the rapid increase of Internet users over the past decade.

In 1958, an obscure Department of Defense agency by the name of the Advanced Research Projects Agency (ARPA), was created in response to the surprise Soviet launch of the *Sputnik 1* earth-orbiting satellite on October 4, 1957—beating out similar US efforts. Suddenly, the United States was playing also-ran to the Soviets, a position that the American public and our government and military leaders were not comfortable with. Renamed DARPA— an acronym standing for Defense Advanced Research Projects Agency—in 1972, the agency's mission was to prevent future technological surprises. For the most part, it has performed this mission well, while generating remarkable new technologies for our military services. Some of the most notable of these innovations include Global Positioning Satellites (GPS), stealth aircraft technology, gallium arsenide-based semiconductors, and the Internet.

The Internet that we know today is a direct descendent of an electronic computer network by the name of ARPANET. When the Soviets launched Sputnik in 1957, they proved that they had a rocket powerful enough to lob a heavy object into space. Before Sputnik, a Soviet nuclear attack would have to be launched by heavy bombers, which would take hours to travel from their Soviet bases to targets in the United States, providing American military officers plenty of time to run their calculations, contact higher authorities (including the president), and make decisions and plans for mounting an effective air defense.

After Sputnik, however, the Soviets could put a nuclear bomb on top of this intercontinental missile and deliver it to the United States in just thirty minutes. This dramatically shortened the amount of time available for military commanders to notify the chain of command and make decisions that could very well lead to all-out nuclear war. In addition, given that the possibility of nuclear war was increasing as the Cold War continued to heat up, our nation would need communications systems that would have the best chance of surviving an attack. Not only would the enemy target our communications systems first in an attempt to decapitate the military, but the electromagnetic pulse (EMP) from a high-altitude blast could potentially knock out communications over very wide areas.

Steve Lukasik—a physicist who served as deputy director and director of the ARPA/DARPA from 1967 through 1974 and chief scientist of the Federal Communications Commission from 1979 through 1982—was a member of the ARPA team during the time the Internet was first being developed by the agency. According to Lukasik, "If you really want to know who is responsible for the Internet, it's Joseph Stalin. Trust me—if there wasn't Joseph Stalin, there would never have been an ARPA and there wouldn't have been an ARPANET nor an Internet."[300]

Before the ARPANET could be built, some key pieces of technology had to be invented first. Paul Baran, Leonard Kleinrock, and Donald W. Davies originally devised the first key piece of technology: packet switching. Baran's system—called the Distributed Adaptive Message Block Network—gained the enthusiastic interest of the US Air Force but was quickly shot down by the Office of the Secretary of Defense, which told the air force brass that it would not be allowed to deploy its own communications system. Communications were considered the province of the Defense Communication Agency. Ultimately, this first attempt at a network died—a victim of turf wars.

In the meantime, ARPA had picked up the task of working on command and control for the services. With that task came a surplus Q32 computer, a large IBM mainframe located at the headquarters of Systems Development Corporation (SDC) in Santa Monica, California, that had been used by the air force for its Semi Automatic Ground Environment (SAGE) air defense system.

The mandate from the Office of the Secretary of Defense? To "do something useful with it."

At the time, ARPA's director Jack Ruina didn't have anyone to run this high-powered computer, so in 1962 he hired an MIT psychology and computing expert by the name of J. C. R. Licklider ("Lick" for short) to head up two departments. The first department was in charge of ARPA's command and control program, while the other department was in charge of behavioral sciences. As it turned out, Licklider had a vision about computing that was part computing and part behavioral—how do computers and humans interact with one another? Lick soon brought in universities—including UCLA, Berkeley, and Stanford—and set up research contracts with them to provide the brainpower necessary to find promising new uses for ARPA's newfound computing resources. This core group—dubbed the Intergalactic Computer Network by Lick—would eventually create the ARPANET.

Lick's vision was for people in different places—all of whom had access to computers and databases—to be connected together in a network to solve problems in a distributed way. Different people could work on a problem simultaneously and send partial results to one another in real time, helping to move the solution forward. However, in 1964—before Lick could implement his vision at ARPA—he headed back to MIT to work on computer timesharing under Project MAC.

In 1966, a couple of years after the departure of Lick, Robert Taylor took over his former position as director of ARPA's Information Processing Techniques Office (IPTO). It was Taylor's idea to share expensive computing resources by linking together all of the computers at ARPA-funded institutions. Taylor's pick to manage the project was a young computer scientist by the name of Larry Roberts. Today, Larry Roberts is generally regarded as the principal architect of the ARPANET, although much credit must also be given to the BBN team—which included Bob Kahn, Dave Walden, Will Crowther, and Severo Ornstein—for working out the details.

And Larry Roberts himself cites the work originally done by J. C. R. Licklider as instrumental to the final result:

Lick had this concept of the intergalactic network, which he believed was everybody could use computers anywhere and get at data anywhere in the world. He didn't envision the number of computers we have today by any means, but he had the same concept—all of the stuff linked together throughout the world, that you can use a remote computer, get data from a remote computer, or use lots of computers in your job. The vision was really Lick's originally. None of us can really claim to have seen that before him nor [can] anybody in the world. Lick saw this vision in the early sixties. He didn't have a clue how to build it. He didn't have any idea how to make this happen. But he knew it was important, so he sat down with me and really convinced me that it was important and convinced me into making it happen.[301]

Larry Roberts decided that the original ARPANET would be comprised of four computers (called Interface Message Processors, or IMPs) in disparate locations—one at University of California, Santa Barbara; one at University of California, Los Angeles; one at Stanford Research Institute (SRI); and one at the University of Utah. Each of these four locations brought something unique to the table. The University of Utah specialized in graphics, while UCLA was going to be the measurement maestro for the system (the school had Leonard Kleinrock, another influential computer scientist who had pioneered packet switching in the early 1960s). And SRI had its NLS (oNLine System), an innovative hypermedia groupware system developed by Doug Engelbart that enabled the creation of digital libraries and the storage and retrieval of electronic documents using hypertext.

On September 1, 1969, the first IMP arrived at UCLA, and the second one was delivered to SRI a month later. On October 29, 1969, the first link was established between the two computers, and the ARPANET was born. By the end of the year, all four computers were connected together, and the ARPANET began to grow. During the early 1970s, other independent computer networks began to spring up, but they could not communicate with one another—that is, until Vint Cerf and Bob Kahn came up with another key innovation, Transmission Control Protocol (TCP).

This was the true birth of the Internet, and it quickly grew, as an increasing number of researchers and institutions jumped on board the network. During this time, ARPA became DARPA, and in July 1975, responsibility for ARPANET was turned over to the Defense Communications Agency.

By the early 1980s, the ARPANET had landed on the radar screen of another federal agency—the National Science Foundation (NSF)—which realized that if the network was so good for computer science, it ought to be just as good for physics, chemistry, and other scientific fields. It therefore set out to create its own computer network—first for computer science departments, and then for universities and colleges at large.

About the Authors

Dr. J. Robert Beyster

Dr. J. Robert Beyster founded SAIC in 1969 and served as the company's chairman and CEO until he retired in 2004. As chairman of the Foundation for Enterprise Development (FED; www.fed.org), a nonprofit organization he founded in 1986, he actively promotes entrepreneurial employee ownership nationally to science and technology communities. In addition to supporting entrepreneurs directly in advancing important, disruptive innovations and growing small businesses through government-funded programs, innovation-based competitions, and publications, the FED actively funds and participates in interdisciplinary research and education at universities.

Dr. Beyster was born in Detroit, Michigan, in 1924 and his family moved to Grosse Ile, Michigan, in 1929. After high school, he enlisted in the US Navy and was sent to college at the University of Michigan to attend the V12 Officer Training Program. He was commissioned as an ensign and eventually served on a destroyer based in Norfolk, Virginia, before leaving the service six months later. Dr. Beyster received a BSE in engineering and physics in 1945, an MS in physics in 1947, and a PhD in nuclear physics in 1950—all from the University of Michigan.

He began his career as a senior scientist at Westinghouse, working in the company's atomic power division on the nuclear submarine program. Within a year, Dr. Beyster left to join Los Alamos National Laboratory, where he worked in the physics department, focused on experimental physics, and coauthored publications with Dr. Hans Bethe. He joined General Atomic in 1957 as chairman of the Accelerator Physics Department, where during his twelve years at the company, Dr. Beyster established the 100-million-volt traveling wave linear accelerator facility and grew it to about 130 staff members.

Beyster founded SAIC—Science Applications International Corpora-tion—on February 3, 1969, in La Jolla, California. SAIC provides a wide range of information technology services, including systems engineering and project management, to federal and state agencies, including all branches of the US military and commercial customers. For years, SAIC was the second-largest employee-owned company in the United States. As of early 2013, the com-pany comprised more than 40,000 employees and has grown to annual rev-enues of more than $10 billion.

Dr. Beyster has written or co-authored approximately 60 publications, reports, and books including *The SAIC Solution: How We Built an $8 Billion Employee-Owned Technology Company*, published by John Wiley & Sons in 2007. A new edition of the book will be released in late 2013.

A fellow of the American Nuclear Society, Dr. Beyster has served as Chairman of its Reactor Physics Division and Shielding Division. He is a fellow of the American Physical Society, a member of the Scientific Advisory Group to the Director, Strategic Target Planning Staff of the Joint Chiefs of Staff, and a member of the National Academy of Engineering. He also serves as Chairman Emeritus of the Board of Directors of the University of California San Diego Foundation.

DARPA designate Dr. Beyster an Honorary Program Manager for his dis-tinguished contributions to the agency over his career. He received the Engi-neering Manager of the Year Award in 2000 from the American Society of Engineering Management, the 2001 Spirit of San Diego Award from the San Diego Regional Chamber of Commerce, the Lifetime Achievement Award from Ernst & Young in 2003, and the Supporter of Entrepreneurialism award from Arthur Young and *Venture* magazine.

In 2006, the San Diego Regional Economic Development Corporation (EDC) recognized Dr. Beyster with the Herb Klein Civic Leadership Award, and he is the recipient of a lifetime achievement award from the University of California, San Diego's CONNECT program for providing 25 years of outstand-ing service to the community. The Horatio Alger Association for Distinguished Americans selected Dr. Beyster to be a 2008 Horatio Alger Award recipient. This honor is bestowed upon those individuals who have overcome adversity to achieve great successes through the American free enterprise system.

Dr. Beyster lives in La Jolla, California, with his wife, Betty and enjoys weekly outings on his boat, *Solutions*

Michael A. Daniels

Michael (Mike) Daniels is former chairman of Network Solutions, Inc. (1995–2000), former chairman and CEO of Mobile365 (2005–2006), and former chairman of the Northern Virginia Technology Council (the largest technology council in the United States).

Mr. Daniels rose to national industry prominence with SAIC, where he served as a member of the senior management team (senior vice president and sector vice president) from 1986 to 2004. He sold his company to SAIC in 1986, when SAIC had revenues of $500 million, and helped build SAIC to its current $11 billion in annual revenues. While at SAIC, he identified and purchased Network Solutions for $4.7 million in March 1995. He led Network Solutions (NSI) through a successful initial public offering (IPO) in 1997 and two successful follow-on offerings in 1999 and 2000. In January 2000, Mr. Daniels led Network Solutions in a follow-on offering of $2.3 billion, one of the largest technology equity offerings in history. Five years after he purchased NSI, he was instrumental in the company's sale on SAIC's behalf for $19.3 billion. Network Solutions contributed $3.4 billion in cash to SAIC, one of the largest transactions in SAIC's thirty-seven-year history.

Mr. Daniels began working part-time for SAIC in May 2004 and was asked to join the board of directors of Apogen Technologies in McLean, Virginia, in May 2004. He worked closely with the CEO and executive management team and Arlington Capital Partners of Washington, DC, in the company's roll-up of two firms, and Apogen was eventually built into a $200 million annual revenue business. Apogen was sold in September 2005 to QinetiQ of the UK (London Stock Exchange) for $300 million in cash.

Mr. Daniels was selected in a national executive search to become chairman of the board of Mobile365 in May 2005 and worked closely with the CEO and the executive management team to build the company. He was asked by the board of directors to also assume the role of CEO, and he served as chairman and CEO beginning in December 2005. He worked with the board of directors to reposition the company, and in November 2006, Mobile365 was sold to Sybase (New York Stock Exchange) for $417 million in cash.

In addition to the above, Mr. Daniels has served on the board of directors of Telcordia Technologies (1998–2003). Formerly Bellcore, this company was purchased by SAIC in 1997 for $650 million and sold in 2003 to private equity firms for $1.3 billion. He also served on the board of directors of VeriSign (Nasdaq) from 2000 to 2001 following the sale of Network Solutions to Veri-Sign in 2000.

Mr. Daniels has served with the Defense Advanced Research Projects Agency (DARPA) of the US Department of Defense, on numerous international commissions and advisory boards, on the United States Trade Representatives International Investment Policy Advisory Commission, as a US National Commissioner to UNESCO, as a senior White House advisor on international technology, and as a senior advisor to the National Security Council.

He currently serves on the board of directors of the Northern Virginia Technology Council (NVTC), Mercury Systems, CACI International, Acentia, the Virginia Chamber of Commerce, and the Boy Scouts of America National Capital Area Council. Mr. Daniels also serves as chairman of the board of GlobalLogic, the Logistics Management Institute, and Invincea.

ACKNOWLEDGMENTS

Writing this book was truly a labor of love. The beginnings, growth, acquisition, and subsequent sale of Network Solutions, Inc.—and SAIC's involvement in it—is an important story that I believe has neither been told in any detail nor with any great degree of accuracy. What little material exists on the topic is often conflicting, inconsistent, or just plain wrong. My primary goal throughout the process was to tell an accurate story of Network Solutions by collecting as many first-person sources as possible through personal interviews and extensive searches of contemporary newspaper and magazine articles and books quoting key people involved in NSI's story. In some cases these people worked for NSI; in other cases people outside the company played pivotal roles in its startup and growth.

I would first like to thank my coauthor Mike Daniels for all his help with this book since we began working on it in early 2008. It was Mike who originally identified the opportunity that this small, 8(a) minority-owned technology business potentially offered to SAIC, and he closely tracked NSI for several years before pushing us to acquire the company. It was primarily through Mike's persistence in pushing this opportunity that we did eventually decide to make a play to acquire NSI. After the acquisition, Mike initially guided the company's growth, serving as NSI's chairman and for some time as its CEO. Mike contributed his own extensive insights to this book, and he set up and conducted interviews with many of the key people quoted in this book. In addition, his review and feedback of the manuscript at numerous critical junctures were vital and extremely helpful.

I would also like to thank all of the many men and women who graciously agreed to be interviewed for the book. Without their thoughts and words, this book would not exist. Gary Desler—the original founder of Network Solutions, before it was incorporated in 1979—provided a treasure trove of previously undocumented details about the founding of the company, the unique technology niche that NSI carved out as it grew, and the challenges the company faced before its acquisition by SAIC. In addition, Ty Grigsby—who with Gary Desler, Emmit McHenry, and Ed Peters, incorporated the business in 1979—provided important and illuminating information and perspectives on the company's early days and its subsequent growth and sale to SAIC. He also provided original company documents (reproduced in this book within appendix 1) that I imagine have not been seen by anyone—in or out of NSI—for decades.

Gabe Battista became CEO of Network Solutions on November 1, 1996, and he led the company's initial public offering on the NASDAQ in 1997. Gabe provided us with firsthand insights into the unique governance issues that NSI faced as it rode the wave of exponential growth in the domain name registry. Vint Cerf is widely considered to be one of the fathers of the Internet, and as a program manager at the Defense Advanced Research Projects Agency (DARPA), he is credited (along with Bob Kahn) with developing the TCP/IP technology that is the heart of the Net's basic communication protocols. Vint offered his perspectives on the origins and growth of the Internet, the technology that makes it run, and the future of this essential worldwide communications network.

Esther Dyson is a commentator and investor in Internet and emerging digital technology. Esther was founding chairman of ICANN (the Internet Corporation for Assigned Names and Numbers), and she served as chairman of the Electronic Frontier Foundation. Esther provided her stimulating and often contrarian points of view on a variety of topics related to the Internet past, present, and future.

David Holtzman was chief technology officer of Network Solutions and manager of the Internet's master root server (the A server) beginning in 1996. David contributed his views on the nature of innovation at NSI, the night the Internet died, and the challenges of keeping up with the Internet's tremendous surge in growth. Bob Kahn, who is credited along with Vint Cerf for developing the TCP/IP protocols, is an Internet pioneer who, while at Bolt Beranek and Newman, was responsible for the system design of the Arpanet, the first packet-switched network. Bob offered his first-hand perspectives on the evolution of the Internet and its transition from what was originally a government communications network into the business powerhouse that it is today.

Jay Killeen joined SAIC in 1994, where he was assigned the task of creating a government affairs department. He soon found his hands full after SAIC acquired Network Solutions, and his insights into the process that he and his team put into place to educate and inform members of Congress about the importance of the Internet and NSI's critical role in ensuring its orderly growth were illuminating. As NSI's CFO, and as a part of Mike Daniels' SAIC team before that, Bob Korzeniewski played major roles in orchestrating SAIC's acquisition of NSI, NSI's successful 1997 initial public offering, and the ultimate sale of the company to VeriSign in 2000. Bob contributed to this book an insider's perspective on the financial side of these complex transactions, as well as the intense litigation environment that NSI endured for much of the time it was owned by SAIC.

Mark Kosters joined Network Solutions, Inc., in July 1991, soon after the company was awarded a subcontract to run the Defense Data Network Network Information Center (DDN NIC, previously operated by SRI International), providing much of the technical expertise required to grow the domain name registry and keep it up and running 24/7. Mark discussed the technical challenges that NSI faced during the 1990s and the pressure the company faced to split the registrar and registry functions of the domain name system. Steve Lukasik is a former director of the Advanced Research Projects Agency (now DARPA) and former chief scientist of the Federal Communications Commission. Steve's personal knowledge of the history of ARPANET and the current state of Internet security is extensive—and an important contribution to this book.

Bill Roper served as a senior vice president and chief financial officer of SAIC from 1990 to 2000, a member of VeriSign's board of directors from 2003 to 2008, and president and CEO of VeriSign from 2007 to 2008. Bill's insights into SAIC's contributions to NSI's growth, and the NSI initial public offering and subsequent stock transactions—and eventual acquisition by VeriSign— were invaluable. Phil Sbarbaro served as outside counsel for Network Solutions, and he was personally involved in defending the company from at least one hundred lawsuits—almost all of which NSI won. Phil brought his inside stories of some of the more interesting cases to this book, which I believe have not been told anywhere else.

In his many years at the National Science Foundation, George Strawn led the transition of the Internet from its military and academic origins to the global business powerhouse that it is today. George contributed his vast knowledge of the technical side of this transition, as well as his experience on the political side of the equation. Don Telage served a variety of roles at NSI, including the company's first president, its chief operating officer for a time, and senior vice president and director. In addition, Don is a mathematician and engineer who played a critical role in turning around NSI after SAIC acquired the company, and he recounts many of his experiences in great detail in this book.

I would also like to express my sincere and deep appreciation to the Red Team for this project, the members of which dedicated a significant amount of their own time and energy to ensure that this book is as accurate as humanly possible and that the overall narrative makes business and technical sense. The red team included Michael Bailey, Mary Ann Beyster, Vint Cerf, Mike Daniels, David Holtzman, Jay Killeen, Mark Kosters, Steve Lukasik, Bill Roper, Jim Russell, and George Strawn.

My daughter Mary Ann Beyster, president of the Foundation for Enterprise Development (www.fed.org) provided ongoing support to this project as well as a critical eye and valuable opinion at numerous junctures. Without Mary Ann's support and guidance, this project would have died within a couple months after we decided to move forward. Peter Economy (www.petereconomy.com)—my coauthor on *The SAIC Solution: How We Built an $8 Billion Employee-Owned Technology Company*—tirelessly helped write, edit, research this book, and update it over the five years taken from concept to publishing, and he participated in the interview process. Wai-Lean Roos, Bianca Lipshitz, and Jamie Dickerson at the Foundation for Enterprise Development provided administrative support to the project, helping to organize our many interviews, create figures, work with our publisher, and ensure that permissions and other legal documents were completed and in place.

Last but not least, I would like to thank my personal assistant, Ralph Callaway. For more than seventeen years now, Ralph has helped to keep my busy life organized, and I am forever grateful for his assistance. Ralph helped to organize my October 2008 trip to Washington, DC, and conduct interviews for this book, and he was instrumental in setting up the interview schedule and keeping our appointments on track.

Thanks to each one of you for the critical roles you played in turning this book from notion to reality. As we so often realized during the five short years that we owned Network Solutions, none of us is as smart as all of us.

NOTES

[1] U.S. Securities and Exchange Commission, "VeriSign Inc. SEC Form 8-K/A, Commission File No. 000-23593," June 8, 2000.

[2] Matthew Zook, "History of gTLD Domain Name Growth" *Zooknic Internet Intelligence*, http://www.zooknic.com/Domains/counts.html.

[3] John Sununu, "The Postal Service's Ticking Time Bomb," *The Boston Globe*, September 26, 2011, http://www.boston.com/bostonglobe/editorial_opinion/oped/articles/2011/09/26/the_postal_services_ticking_time_bomb/.

[4] J. Robert Beyster with Peter Economy, *The SAIC Solution: How We Built an $8 Billion Employee-Owned Technology Company* (Hoboken, NJ: John Wiley and Sons, 2007), 93.

[5] "Executive Team," *Nominum*, http://www.nominum.com/company/about/leadership/.

[6] "Internet Protocol Suite," *Wikipedia*, last modified August 8, 2013, http://en.wikipedia.org/wiki/Internet_Protocol_Suite.

[7] Vinton Cerf, Yogen Dalal, and Carl Sunshine, "Specification of Internet Transmission Control Program," December 1974, http://tools.ietf.org/html/rfc675.

[8] "WHOIS Behind That Domain?" *Network Solutions*, http://www.networksolutions.com/whois/index.jsp.

[9] Ross Wm. Rader, "One History of DNS," April 25, 2001, 3, http://www.byte.org/one-history-of-dns.pdf.

[10] D.L. Mills, "RFC 799—Internet Name Domains," September 1981, http://www.faqs.org/rfcs/rfc799.html.

[11] "IANA—Internet Assigned Numbers Authority," http://www.livinginternet.com/i/iw_mgmt_iana.htm.

[12] "Postel Disputes: The Internet," *The Economist*, February 8, 1997.

[13] "Port (Computer Networking)," *Wikipedia*, last modified August 7, 2013, http://en.wikipedia.org/wiki/Port_number.

[14] Kim Davies, "An Introduction to IANA," (presentation notes), September 29, 2008, http://www.iana.org/about/presentations/davies-atlarge-iana101-paper-080929-en.pdf.

[15] *SRI Alumni Association Newsletter*, April 2011, http://www.sri.com/sites/default/files/brochures/apr-11.pdf.

¹⁶ Zaw-Sing Su and Jon Postel, "RFC 819—Domain Naming Convention for Internet User Applications," August 1982, http://www.faqs.org/rfcs/rfc819.html.

¹⁷ Paul Mockapetris and Kevin Dunlap, "Development of the Domain Name System," *ACM SIGCOMM Computer Communication Review*, 1988, 123.

¹⁸ Paul Mockapetris, "RFC 883—Domain Names: Implementation Specification," November 1983, http://www.faqs.org/rfcs/rfc883.html.

¹⁹ Jon Postel, "RFC 1591—Domain Name System Structure and Delegation," March 1994, http://www.faqs.org/rfcs/rfc1591.html.

²⁰ "100 Oldest .COM Domains," *iwhois.com.*, https://www.iwhois.com/oldest/.

²¹ "Paul V. Mockapetris, Ph.D. Chief Scientist & Chairman Nominum," (interview) *Who is Who in the Internet World*, http://www.wiwiw.org/test/index.php/wiwiw/pioneers/paul_v_mockapetris.

²² "Root Name Server," *Wikipedia*, last modified August 6, 2013, http://en.wikipedia.org/wiki/Root_nameserver.

²³ Daniel Karrenberg, "The Internet Domain Name System Explained for Non-Experts" (briefing paper), *Internet Society*, March 1, 2004, http://www.isoc.org/briefings/019/.

²⁴ *Root-servers.org*, http://www.root-servers.org/.

²⁵ Daniel Karrenberg, "The Internet Domain Name System Explained for Non-Experts" (briefing paper), *Internet Society*, March 1, 2004, http://www.isoc.org/briefings/019/.

²⁶ Mark Kosters, Red Team Review (notes), 2011.

²⁷ George Strawn, interview by J. Robert Beyster and Mike Daniels, September 16, 2008.

²⁸ "NSF at a Glance," *National Science Foundation*, http://nsf.gov/about/glance.jsp.

²⁹ Vinton Cerf, Red Team Review (notes), 2011.

³⁰ "NSFNET Backbone Service Restructured," *Link Letter*, November 1992, NSFNET Information Services, Merit Network, Inc., Ann Arbor.

³¹ Vinton Cerf, Red Team Review (notes), 2011.

³² George Strawn, interview by J. Robert Beyster and Mike Daniels, September 16, 2008.

³³ "Internet History: NSFNET" (derived from Kevin Werbach, "Digital Tornado: The Internet and Telecommunications Policy," FCC Office of Plans and Policy Working Paper No. 29 [March 1997]: 13), http://www.cybertelecom.org/notes/nsfnet.htm.

³⁴ "Internet History: NSFNET" (derived from Kevin Werbach, "Digital Tornado: The Internet and Telecommunications Policy," FCC Office of Plans and Policy Working Paper No. 29 [March 1997]: 13), http://www.cybertelecom.org/notes/nsfnet.htm.

³⁵ Vinton Cerf, Red Team Review (notes), 2011.

³⁶ "The Launch of NSFNET," *National Science Foundation*, http://www.nsf.gov/about/history/nsf0050/internet/launch.htm.

³⁷ Steve Lukasik, "Bob Beyster Book Comments" (unpublished notes), June 2011.

[38] Vinton Cerf, interview by Mike Daniels, October 31, 2008.

[39] *Scientific and Advanced-Technology Act of 1992.* Public Law 102-476, 102nd Cong., 2nd sess., (October 23, 1992), http://thomas.loc.gov/cgi-bin/bdquery/z?d102:SN01146:@@@ L&summ2=m&.

[40] "42 USC § 1862—Functions," *Cornell University Law School Legal Information Institute,* http://www.law.cornell.edu/uscode/42/usc_sec_42_00001862----000-.html.

[41] Bob Kahn, interview by Peter Economy and Mike Daniels, October 27, 2008.

[42] George Strawn, interview by J. Robert Beyster and Mike Daniels, September 16, 2008.

[43] George Strawn, interview by J. Robert Beyster and Mike Daniels, September 16, 2008.

[44] "National Science Foundation Network," *Wikipedia,* last modified July 3, 2013, http://en.wikipedia.org/wiki/National_Science_Foundation_Network.

[45] "World Wide Web," *Wikipedia,* last modified August 10, 2013, http://en.wikipedia.org/wiki/World_Wide_Web.

[46] John Borland, "Browser Wars: High Price, Huge Rewards," *ZDNet,* April 15, 2003, http://www.zdnet.com/news/browser-wars-high-price-huge-rewards/128738.

[47] Cliff Hocker, "Graduating with Honors," *Black Enterprise,* November 1990, 88–94.

[48] Gary Desler, interview by Peter Economy, May 26, 2010.

[49] Gary Desler, interview by Peter Economy, May 26, 2010.

[50] Gary Desler, interview by Peter Economy, May 26, 2010.

[51] Gary Desler, interview by Peter Economy, May 26, 2010.

[52] Gary Desler, interview by Peter Economy, May 26, 2010.

[53] Tyrone W. Grigsby, biographical sketch, courtesy of Mr. Grigsby, August 16, 2010.

[54] Tyrone Grigsby, interview by Peter Economy, August 18, 2010.

[55] Tyrone Grigsby, interview by Peter Economy, August 18, 2010.

[56] Gina Henderson, "debacle.com," *Emerge,* May 2000, 137–38.

[57] Tyrone Grigsby, interview by Peter Economy, August 18, 2010.

[58] Gary Desler, interview by Peter Economy, May 26, 2010.

[59] Tyrone Grigsby, interview by Peter Economy, August 18, 2010.

[60] Gary Desler, interview by Peter Economy, May 26, 2010.

[61] Tyrone Grigsby, interview by Peter Economy, August 18, 2010.

[62] Gary Desler, interview by Peter Economy, May 26, 2010.

[63] Tyrone Grigsby interview for *Nothing But a Jar of Oil (Seven Steps to Achieving Financial Victory Through Biblical Entrepreneurship),* by Patrice Tsague & Melvin Mooring (Damascus, MD: Nehemiah Publishing, 2009).

[64] Tyrone Grigsby, interview by Peter Economy, August 18, 2010.

[65] Cliff Hocker, "Graduating with Honors," *Black Enterprise,* November 1990, 88–94.

[66] Cliff Hocker, "Graduating with Honors," *Black Enterprise,* November 1990, 88–94.

[67] U.S. Department of Agriculture, "Section 8(a)," *Office of Small and Disadvantaged Business Utilization*, http://www.dm.usda.gov/smallbus/8a.htm.

[68] Tyrone Grigsby, interview by Peter Economy, August 18, 2010.

[69] Tyrone Grigsby, interview by Peter Economy, August 18, 2010.

[70] Tyrone Grigsby, interview by Peter Economy, August 18, 2010.

[71] Tyrone Grigsby, interview by Peter Economy, August 18, 2010.

[72] Tyrone Grigsby, interview by Peter Economy, August 18, 2010.

[73] "40th Anniversary of the Internet," *UCLA Engineering*, October 29, 2009, http://www.engineer.ucla.edu/IA40/.

[74] Tyrone Grigsby, interview by Peter Economy, August 18, 2010.

[75] "NSI 3-Year Plan, Chapter 1: Statement and Goals and Objectives," (unpublished manuscript, courtesy of Tyrone Grigsby), 1993.

[76] Cliff Hocker, "Graduating with Honors," *Black Enterprise*, November 1990, 88–94.

[77] Tyrone Grigsby, interview by Peter Economy, August 18, 2010.

[78] Gary Desler, interview by Peter Economy, May 26, 2010.

[79] Tyrone Grigsby, interview by Peter Economy, August 18, 2010.

[80] Tyrone Grigsby, interview by Peter Economy, August 18, 2010.

[81] Cliff Hocker, "Graduating with Honors," *Black Enterprise*, November 1990, 88–94.

[82] "InterNIC," *Wikipedia*, last modified April 9, 2013, http://en.wikipedia.org/wiki/InterNIC.

[83] S. Williamson and L. Nobile, "RFC 1261—Transition of NIC Services," September 1991, http://www.faqs.org/rfcs/rfc1261.html.

[84] S. Williamson and L. Nobile, "RFC 1261—Transition of NIC Services," September 1991, http://www.faqs.org/rfcs/rfc1261.html.

[85] Mark Kosters, "Mark Kosters, Chief Technology Officer" (LinkedIn profile of Mark Kosters), http://www.linkedin.com/profile/view?id=6250393&authType=name&authToken=zrn4&pvs=pp&trk=ppro_viewmore.

[86] Mark Kosters, interview by Mike Daniels, October 21, 2008.

[87] "NSF Cooperative Agreement NCR-9218742" (parties: National Science Foundation and Network Solutions Incorporated), January 1, 1993.

[88] Mark Kosters, interview by Mike Daniels, October 21, 2008.

[89] "NSF Cooperative Agreement NCR-9218742" (parties: National Science Foundation and Network Solutions Incorporated), January 1, 1993.

[90] Mark Kosters, Red Team Review (notes), 2011.

[91] "NSF Cooperative Agreement NCR-9218742" (parties: National Science Foundation and Network Solutions Incorporated), January 1, 1993.

[92] "NSF Cooperative Agreement NCR-9218742" (parties: National Science Foundation and Network Solutions Incorporated), January 1, 1993.

[93] George Strawn, interview by J. Robert Beyster and Mike Daniels, September 16, 2008.

[94] "NSF Cooperative Agreement NCR-9218742" (parties: National Science Foundation and Network Solutions Incorporated), January 1, 1993.

[95] "NSF Cooperative Agreement NCR-9218742" (parties: National Science Foundation and Network Solutions Incorporated), January 1, 1993.

[96] George Strawn, interview by J. Robert Beyster and Mike Daniels, September 16, 2008.

[97] George Strawn, interview by J. Robert Beyster and Mike Daniels, September 16, 2008.

[98] "Board" (Onramp Wireless board members), *Onramp Wireless*, http://www.onrampwireless.com/about-us/board/.

[99] Kelly Macavinta, "How .com Became .$$$," *CNETnews.com*, February 15, 1999, http://news.cnet.com/2009-1082-221582.html.

[100] "NSF Cooperative Agreement NCR-9218742" (parties: National Science Foundation and Network Solutions Incorporated), January 1, 1993.

[101] Gary Desler, interview by Peter Economy, May 26, 2010.

[102] Mark Kosters, Red Team Review (notes), 2011.

[103] Tyrone Grigsby, interview by Peter Economy, August 18, 2010.

[104] Tyrone Grigsby, interview by Peter Economy, August 18, 2010.

[105] Gary Desler, interview by Peter Economy, May 26, 2010.

[106] "Frequently Asked Questions," *SBA Office of Advocacy*, January 2011, http://www.sba.gov/sites/default/files/sbfaq.pdf.

[107] Jay Goltz, "Top 10 Reasons Small Businesses Fail," Business Day Small Business, *New York Times*, January 5, 2011, http://boss.blogs.nytimes.com/2011/01/05/top-10-reasons-small-businesses-fail/.

[108] Mark Kosters, interview by Mike Daniels, October 21, 2008.

[109] Tyrone Grigsby, interview by Peter Economy, August 18, 2010.

[110] Gary Desler, interview by Peter Economy, May 26, 2010.

[111] Gina Henderson, "debacle.com," *Emerge*, May 2000.

[112] Tyrone Grigsby, interview by Peter Economy, August 18, 2010.

[113] Tyrone Grigsby, interview by Peter Economy, August 18, 2010.

[114] Gina Henderson, "debacle.com," *Emerge*, May 2000.

[115] Gina Henderson, "debacle.com," *Emerge*, May 2000.

[116] Gina Henderson, "debacle.com," *Emerge*, May 2000.

[117] Gary Desler, interview by Peter Economy, May 26, 2010.

[118] Mark Kosters, interview by Mike Daniels, October 21, 2008.

119 Glenn Simpson and John Simons, "The Dotted Line: A Little Internet Firm Got a Big Monopoly; Is That Such a Bad Thing?" *Wall Street Journal*, sec. A1, October 8, 1998.

120 Mark Kosters, interview by Mike Daniels, October 21, 2008.

121 Bill Roper, interview by J. Robert Beyster, October 6, 2008.

122 Mike Daniels, interview by J. Robert Beyster, September 18, 2008.

123 Bruce Bigelow, "The Untold Story of SAIC, Network Solutions, and the Rise of the Web— Part 1," *Xconomy*, July 29, 2009, http://www.xconomy.com/san-diego/2009/07/29/the-untold-story-of-saic-network-solutions-and-the-rise-of-the-web-part-1/.

124 Mike Daniels, interview by J. Robert Beyster, September 18, 2008.

125 Mike Daniels, interview by J. Robert Beyster, September 18, 2008.

126 Mike Daniels, interview by J. Robert Beyster, September 18, 2008.

127 Bob Korzeniewski, interview by J. Robert Beyster and Mike Daniels, September 17, 2008.

128 Mike Daniels, interview by J. Robert Beyster, September 18, 2008.

129 Mike Daniels, interview by J. Robert Beyster, September 18, 2008.

130 Bob Korzeniewski, interview by J. Robert Beyster and Mike Daniels, September 17, 2008.

131 "Emmit J. McHenry," *Aetna*, 1995, http://www.aetna.com/foundation/aahcalendar/1995mchenry.html.

132 Gary Desler, interview by Peter Economy, May 26, 2010.

133 Patrice Tsague, *Biblical Principles for Starting and Operating a Business*, (Bloomington, IN: AuthorHouse, 2006), xviii.

134 Don Telage, interview by Mike Daniels, November 4, 2008.

135 "NSF Sensational 60," *National Science Foundation*, 2010, http://www.nsf.gov/about/history/sensational60.pdf.

136 Nick Wingfield, "NSF Ends Internet Subsidy; Domain Names to Cost $50," *InfoWorld*, September 18, 1995, 8.

137 George Strawn, interview by J. Robert Beyster and Mike Daniels, September 16, 2008.

138 Mark Kosters, interview by Mike Daniels, October 21, 2008.

139 Gary Desler, interview by Peter Economy, May 26, 2010.

140 "NSF Cooperative Agreement NCR-9218742, Amendment 4" (parties: National Science Foundation and Network Solutions Incorporated), September 13, 1995.

141 George Strawn, interview by J. Robert Beyster and Mike Daniels, September 16, 2008.

142 Don Telage, interview by Mike Daniels, November 4, 2008.

143 George Strawn, interview by J. Robert Beyster and Mike Daniels, September 16, 2008.

144 Vinton Cerf, interview by Mike Daniels, October 31, 2008.

145 Don Telage, interview by Mike Daniels, November 4, 2008.

146 Glenn Simpson and John Simons, "The Dotted Line: A Little Internet Firm Got a Big Monopoly; Is That Such a Bad Thing?" *Wall Street Journal*, sec. A1, October 8, 1998.

147 "Memorandum of Grounds for Decision," *Science Applications International Corporation v. Comptroller of the Treasury*, Maryland Court, No. 04-IN-OO-0632, May 11, 2006, 5.

148 Mark Kosters, Red Team Review (notes), 2011.

149 Mark Kosters, Red Team Review (notes), 2011.

150 Mark Kosters, interview by Mike Daniels, October 21, 2008.

151 Robin Murphy, "The InterNIC Year in Review," *InterNIC News*, January 1997.

152 Mark Leibovich, "CEO of Network Solutions to Depart," *Washington Post*, November 17, 1998.

153 "Board of Directors," TEOCO, http://www.teoco.com/board-of-directors.

154 Gabe Battista, interview by Mike Daniels, September 17, 2008.

155 Gabe Battista, interview by Mike Daniels, September 17, 2008.

156 "David Holtzman Biography," *David Holtzman*, http://www.globalpov.com/biography.html.

157 "David H. Holtzman," *Wikipedia*, last modified September 14, 2011, http://en.wikipedia.org/wiki/David_H._Holtzman.

158 Mark Kosters, interview by Mike Daniels, October 21, 2008.

159 Mark Kosters, interview by Mike Daniels, October 21, 2008.

160 Mark Kosters, Red Team Review (notes), 2011.

161 David Hilzenrath, "Microsoft-NBC Web Site Wrongly Deactivated in Domain Purge," *Washington Post*, July 2, 1996, C1.

162 Mark Kosters, interview by Mike Daniels, October 21, 2008.

163 Mark Kosters, interview by Mike Daniels, October 21, 2008.

164 Bob Korzeniewski, interview by J. Robert Beyster and Mike Daniels, September 17, 2008.

165 Don Telage, interview by Mike Daniels, November 4, 2008.

166 Don Telage, interview by Mike Daniels, November 4, 2008.

167 Don Telage, interview by Mike Daniels, November 4, 2008.

168 "Network Solutions Inc/DE form S-1," SecInfo (securities information from the US SEC EDGAR database), July 3, 1997, http://www.secinfo.com/dsvRq.82bt.htm#87q.

169 "A Million Domains, Half Unpaid," *CNET News*, March 11, 1997.

170 "Panelists Resource Biographies 2011 Workshops," *Internet Governance Forum*, http://www.intgovforum.org/cms/index.php/component/chronocontact/?chronoformname=2011ResourceBioView&wspid=337.

171 Mark Kosters, interview by Mike Daniels, October 21, 2008.

172 "i.root-servers.net," *Netnod*, http://www.netnod.se/dns/iroot.

[173] David Holtzman, interview by Mike Daniels, October 14, 2008.

[174] David Holtzman, interview by Mike Daniels, October 14, 2008.

[175] Gabe Battista, interview by Mike Daniels, September 17, 2008.

[176] Kevin Sullivan and Brian Oakes, "Competition Has Arrived; It's Just the Beginning for Network Solutions," Lehman Brothers report, September 8, 1999, 7.

[177] "Sample Business Contracts," *Onecle*, http://contracts.onecle.com/netsol/herndon.lease.1997.05.30.shtml.

[178] George Strawn, interview by J. Robert Beyster and Mike Daniels, September 16, 2008.

[179] Tyrone Grigsby, interview by Peter Economy, August 18, 2010.

[180] David Holtzman, interview by Mike Daniels, October 14, 2008.

[181] David Holtzman, interview by Mike Daniels, October 14, 2008.

[182] John Markoff, "Ignored Warning Leads to Chaos on the Internet," *New York Times*, July 18, 1997.

[183] Rajiv Chandrasekaran and Elizabeth Corcoran, "Human Errors Block E-Mail, Web Sites in Internet Failure; Garbled Address Files from Va. Firm Blamed," *Washington Post*, sec. A1, July 18, 1997.

[184] Peter Wayner, "Internet Glitch Reveals System's Pervasiveness, Vulnerability," *New York Times*, July 18, 1997.

[185] John Markoff, "Ignored Warning Leads to Chaos on the Internet," *New York Times*, July 18, 1997.

[186] Mark Kosters, interview by Mike Daniels, October 21, 2008.

[187] Mark Kosters, interview by Mike Daniels, October 21, 2008.

[188] David Holtzman, interview by Mike Daniels, October 14, 2008.

[189] Gabe Battista, interview by Mike Daniels, September 17, 2008.

[190] Gabe Battista, interview by Mike Daniels, September 17, 2008.

[191] Gabe Battista, interview by Mike Daniels, September 17, 2008.

[192] Rajiv Chandrasekaran and Elizabeth Corcoran, "Human Errors Block E-Mail, Web Sites in Internet Failure; Garbled Address Files from Va. Firm Blamed," *Washington Post*, sec. A1, July 18, 1997.

[193] David Holtzman, interview by Mike Daniels, October 14, 2008.

[194] Suzanne Galante, "Network Solutions Nets Strong Trade," *CNET News*, September 26, 1997.

[195] "Netscape," *Wikipedia*, last modified July 31, 2013, http://en.wikipedia.org/wiki/Netscape.

[196] Mike Daniels, interview by J. Robert Beyster, September 18, 2008.

[197] Mike Daniels, interview by J. Robert Beyster, September 18, 2008.

[198] The White House Office of the Press Secretary, "Memorandum for the Heads of Executive Departments and Agencies; Subject: Electronic Commerce," memorandum, July 1, 1997, http://www.fas.org/irp/offdocs/pdd-nec-ec.htm.

[199] George Strawn, interview by J. Robert Beyster and Mike Daniels, September 16, 2008.

[200] Mike Daniels, interview by J. Robert Beyster, September 18, 2008.

[201] Bill Roper, interview by J. Robert Beyster and Mike Daniels, October 6, 2008.

[202] David Holtzman, interview by Mike Daniels, October 14, 2008.

[203] Bill Roper, interview by J. Robert Beyster and Mike Daniels, October 6, 2008.

[204] President William J. Clinton and Vice President Albert Gore, Jr., *A Framework for Global Electronic Commerce*, report from the United States White House Office, July 1, 1997.

[205] K. Hubbard, M. Kosters, D. Conrad, D. Karrenberg, and J. Postel, "RFC 2050—Internet Registry IP Allocation Guidelines," November 1996, http://www.faqs.org/rfcs/rfc2050.html.

[206] K. Hubbard, M. Kosters, D. Conrad, D. Karrenberg, and J. Postel, "RFC 2050—Internet Registry IP Allocation Guidelines," November 1996, http://www.faqs.org/rfcs/rfc2050.html.

[207] V. Cerf, "IAB Recommended Policy on Distributing Internet Identifier Assignment," August 1990, http://www.rfc-editor.org/rfc/rfc1174.txt.

[208] E. Gerich, "Guidelines for Management of IP Address Space," October 1992, http://www.rfc-editor.org/rfc/rfc1366.txt.

[209] "History of APNIC," *APNIC*, http://www.apnic.net/about-APNIC/organization/history-of-apnic.

[210] "History of the Regional Internet Registries," *APNIC*, http://www.apnic.net/about-APNIC/organization/history-of-apnic/history-of-the-regional-internet-registries.

[211] K. Hubbard, M. Kosters, D. Conrad, D. Karrenberg, and J. Postel, "RFC 2050—Internet Registry IP Allocation Guidelines," November 1996, http://www.faqs.org/rfcs/rfc2050.html.

[212] Daniel Karrenberg, Gerard Ross, Paul Wilson & Leslie Nobile, "Development of the Regional Internet Registry System," *The Internet Protocol Journal*, December 2001.

[213] George Strawn, "NSI Book Review," e-mail message to Mary Ann Beyster, April 24, 2012.

[214] Phil Sbarbaro, interview by Mike Daniels and J. Robert Beyster, September 18, 2008.

[215] Don Telage, interview by Mike Daniels, November 4, 2008.

[216] Phil Sbarbaro, review of "Names, Numbers, and Network Solutions," (manuscript), March 12, 2013.

[217] "Articles of Incorporation of American Registry for Internet Numbers, LTD.," *American Registry for Internet Numbers*, August 7, 1997, https://www.arin.net/about_us/corp_docs/artic_incorp.html.

[218] Ross Wm. Rader, "One History of DNS," April 25, 2001, 19-20, http://www.byte.org/one-history-of-dns.pdf.

[219] Craig Simon, email exchange with Jon Postel and commentary, November 12, 2002, http://mailman.postel.org/pipermail/internet-history/2002-November/000376.html.

[220] Katie Hafner and Matthew Lyon, *Where Wizards Stay Up Late: The Origins of the Internet* (New York: Simon and Schuster, 1996).

[221] "Jon Postel," *Wikipedia*, last modified August 11, 2013, http://en.wikipedia.org/wiki/Jon_Postel.

[222] Craig Simon, email exchange with Jon Postel and commentary, November 12, 2002, http://mailman.postel.org/pipermail/internet-history/2002-November/000376.html.

[223] Mark Kosters, interview by Mike Daniels, October 21, 2008.

[224] Jack Goldsmith and Tim Wu, *Who Controls the Internet?* (New York: Oxford University Press, 2006), 44.

[225] Milton Mueller, *Ruling the Root* (Cambridge, MA: MIT Press, 2002), 161–62.

[226] Jack Goldsmith and Tim Wu, *Who Controls the Internet?* (New York: Oxford University Press, 2006), 46.

[227] Mark Kosters, interview by Mike Daniels, October 21, 2008.

[228] Jay Killeen, interview by J. Robert Beyster and Mike Daniels, September 16, 2008.

[229] U.S. Department of Commerce, National Telecommunications and Information Administration, *A Proposal to Improve the Internet Technical Management of Internet Names and Addresses*, January 30, 1998.

[230] J. Postel, "New Registries and the Delegation of International Top Level Domains" (Internet draft), May 1996, http://tools.ietf.org/html/draft-postel-iana-itld-admin-00.

[231] J. Postel, "New Registries and the Delegation of International Top Level Domains" (Internet draft), May 1996, http://tools.ietf.org/html/draft-postel-iana-itld-admin-00.

[232] J. Postel, "New Registries and the Delegation of International Top Level Domains" (Internet draft), May 1996, http://tools.ietf.org/html/draft-postel-iana-itld-admin-00.

[233] J. Postel, "New Registries and the Delegation of International Top Level Domains" (Internet draft), May 1996, http://tools.ietf.org/html/draft-postel-iana-itld-admin-00.

[234] Board of Trustees of Internet Society, "Board Meeting No. 9—Minutes" *Internet Society*, June 24-25, 1996, http://www.isoc.org/isoc/general/trustees/meetings.php?id=9&doc=m.

[235] Ross Wm. Rader, "One History of DNS," April 25, 2001, 15, http://www.byte.org/one-history-of-dns.pdf.

[236] "The Generic Top Level Domain Memorandum of Understanding, Frequently Asked Questions (FAQ)," last updated November 13, 1997, http://web.archive.org/web/19971211190139/www.gtld-mou.org/docs/faq.html#whatis.

[237] "The Generic Top Level Domain Memorandum of Understanding," last updated December 8, 1997, http://web.archive.org/web/19971211190034/http://www.gtld-mou.org/.

[238] John Fuller, "How the International Ad Hoc Committee Worked," *HowStuffWorks*, http://computer.howstuffworks.com/iahc2.htm.

[239] U.S. Department of Commerce, "Improvement of Technical Management of Internet Names and Addresses; Proposed Rule," *Federal Register* 63, no. 34 (February 20, 1998), http://www.ntia.doc.gov/ntiahome/domainname/dnsdrft.htm.

[240] U.S. Department of Commerce, "Improvement of Technical Management of Internet Names and Addresses; Proposed Rule," *Federal Register* 63, no. 34 (February 20, 1998), http://www.ntia.doc.gov/ntiahome/domainname/dnsdrft.htm.

[241] U.S. Department of Commerce, "Improvement of Technical Management of Internet Names and Addresses; Proposed Rule," *Federal Register* 63, no. 34 (February 20, 1998), http://www.ntia.doc.gov/ntiahome/domainname/dnsdrft.htm.

[242] Jay Killeen, interview by J. Robert Beyster and Mike Daniels, September 16, 2008.

[243] Esther Dyson, interview by Peter Economy, December 12, 2008.

[244] Wolfgang Kleinwachter, "From Self-Governance to Public-Private Partnership: The Changing Role of Governments in the Management of the Internet's Core Resources," *Loyola of Los Angeles Law Review* 36, no. 3 (2003), http://digitalcommons.lmu.edu/cgi/viewcontent.cgi?article=2370&context=llr.

[245] Wolfgang Kleinwachter, "From Self-Governance to Public-Private Partnership: The Changing Role of Governments in the Management of the Internet's Core Resources," *Loyola of Los Angeles Law Review* 36, no. 3 (2003), http://digitalcommons.lmu.edu/cgi/viewcontent.cgi?article=2370&context=llr.

[246] "In Memoriam, Dr. Jonathan B. Postel, August 3, 1943 - October 16, 1998," *The Domain Name Handbook*, 1998, http://www.domainhandbook.com/postel.html.

[247] Wolfgang Kleinwachter, "From Self-Governance to Public-Private Partnership: The Changing Role of Governments in the Management of the Internet's Core Resources," *Loyola of Los Angeles Law Review* 36, no. 3 (2003), http://llr.lls.edu/volumes/v36-issue3/kleinwaechter.pdf.

[248] Esther Dyson, interview by Peter Economy, December 12, 2008.

[249] Mark Kosters, interview by Mike Daniels, October 21, 2008.

[250] Mark Kosters, interview by Mike Daniels, October 21, 2008.

251 "NSF Cooperative Agreement NCR-9218742, Amendment 11" (parties: National Science Foundation and Network Solutions Incorporated), October 7, 1998, http://www.icann.org/en/nsi/coopagmt-amend11-07oct98.htm.

252 Gabe Battista, interview by Mike Daniels, September 17, 2008.

253 Phil Sbarbaro, interview by Mike Daniels and J. Robert Beyster, September 18, 2008.

254 "Press Release: ICANN Names Competitive Domain-Name Registrars," *Internet Corporation for Assigned Names and Numbers (ICANN)*, April 21, 1999, http://www.icann.org/en/announcements/icann-pr21apr99.htm.

255 U.S. Securities and Exchange Commission, "Network Solutions, Inc., Form 10-Q," August 3, 1999, http://google.brand.edgar-online.com/EFX_dll/EDGARpro.dll?FetchFilingHTML1?ID=1125890&SessionID=rol6Wv7XX3Cghw7.

256 Network Solutions, "Network Solutions Announces Register.com as First Testbed Registrar to Register Names," press release, June 7, 1999.

257 David Holtzman, interview by Mike Daniels, October 14, 2008.

258 David Holtzman comments on manuscript, 2012.

259 Gabe Battista, interview by Mike Daniels, September 17, 2008.

260 Melanie Austria Farmer, "VeriSign Buys Network Solutions in $21 Billion Deal," *CNET News*, March 7, 2000.

261 Melanie Austria Farmer, "VeriSign Buys Network Solutions in $21 Billion Deal," *CNET News*, March 7, 2000.

262 "VeriSign Buys Domain Firm," *CNNMoney*, March 7, 2000.

263 Michael Lewis, *The New New Thing* (New York: W.W. Norton & Company, 1999), 13–14.

264 Yordanka Bahchevanska "SAIC IPO Raises $1.2 Billion," *IPO Outlook*, October 14, 2006, http://www.123jump.com/ipo-outlook/SAIC-IPO-Raises-$1.2-Billion/19401/.

265 Mark Kosters, interview by Mike Daniels, October 21, 2008.

266 Suzanne Galante, "Network Solutions Nets Strong Trade," *CNET News*, September 26, 1997.

267 Suzanne Galante, "Network Solutions Nets Strong Trade," *CNET News*, September 26, 1997.

268 Suzanne Galante, "Network Solutions Nets Strong Trade," *CNET News*, September 26, 1997.

269 "Memorandum of Grounds for Decision," *Science Applications International Corporation v. Comptroller of the Treasury*, Maryland Court, No. 04-IN-OO-0632, May 11, 2006, 2.

270 Therese Poletti, "A Less Boisterous IPO Mood," *Wall Street Journal*, March 11, 2010.

271 U.S. Securities and Exchange Commission, "VeriSign Inc. SEC Form 425, Commission File No. 000-23593," March 7, 2000.

272 U.S. Securities and Exchange Commission, "VeriSign Inc. SEC Form 8-K/A, Commission File No. 000-23593," June 8, 2000.

273 U.S. Securities and Exchange Commission, "VeriSign Inc. SEC Form 8-K/A, Commission File No. 000-23593," June 8, 2000.

274 Melanie Austria Farmer, "VeriSign Buys Network Solutions in $21 Billion Deal," *CNET News*, March 7, 2000.

275 "VeriSign To Sell Network Solutions, Exit Registrar Business," *TechNewsWorld*, October 17, 2003, http://www.technewsworld.com/story/31890.html?wlc=1275146264.

276 "Reaction to VeriSign-NSI Break Up" (blog), *CircleID*, October 17, 2003, http://www.circleid.com/posts/reaction_to_verisign_nsi_break_up/.

277 Tomio Geron, "Web.com Buying Network Solutions For $560 Million," *Forbes*, August 3, 2011, http://www.forbes.com/sites/tomiogeron/2011/08/03/web-com-buys-network-solutions-for-560-million/.

278 Bill Roper, e-mail message to Peter Economy, June 11, 2011.

279 Bill Flook, "Network Solutions Sold to Web.com for $405M and Stock," *Washington Business Journal*, August 3, 2011, http://www.bizjournals.com/washington/news/2011/08/03/network-solutions-sold-to-webcom.html.

280 Steve Rockwood , interview by J. Robert Beyster, March 16, 2006.

281 Steve Mills, "The Future of Business" (thought leadership paper), *IBM*, June 2007, http://www-935.ibm.com/services/us/cio/pdf/wp-enbusflex_raw11032-usen-00_hr.pdf.

282 "New Generic Top Level Domains: About the Program," *Internet Corporation for Assigned Names and Numbers*, http://newgtlds.icann.org/en/about/program.

283 Association for National Advertisers, "ANA Cites Major Flaws in ICANN's Proposed Top-Level Internet Domain Program" (press release), August 4, 2011, http://www.ana.net/content/show/id/21790.

284 National Restaurant Association, "National Restaurant Association Registers Opposition to New Internet Domain Name Plan" (press release), December 7, 2011, http://www.restaurant.org/Pressroom/Press-Releases/National-Restaurant-Association-Registers-Oppositi.

285 Jason Del Rey, "Senate Calls Hearing on ICANN's Controversial Top-Level Domain Expansion Plan," *Ad Age digital*, December 2, 2011, http://adage.com/article/digital/senate-calls-hearing-icann-s-controversial-top-level-domain-expansion-plan/231350/.

286 "New Generic Top-Level Domains Fact Sheet," *Internet Corporation for Assigned Names and Numbers*, July 2011, http://www.icann.org/en/topics/new-gtlds/gtld-facts-31jul11-en.pdf.

287 "IPv4 Address Exhaustion," *Wikipedia*, last modified April 3, 2013, http://en.wikipedia.org/wiki/IPv4_address_exhaustion.

[288] Bill Roper, interview by J. Robert Beyster, October 6, 2008.

[289] William J. Lynn III, "Defending a New Domain: The Pentagon's Cyberstrategy," *Foreign Affairs*, September/October 2010, http://www.foreignaffairs.com/articles/66552/william-j-lynn-iii/defending-a-new-domain.

[290] "Killer Application," *Wikipedia*, last modified July 19, 2013, http://en.wikipedia.org/wiki/Killer_application.

[291] "Newsroom: Key Facts," *Facebook*, https://newsroom.fb.com/Key-Facts.

[292] "Skype Technologies," *Wikipedia*, last modified August 7, 2013, http://en.wikipedia.org/wiki/Skype_Limited#History.

[293] Chelsi Nakano, "New Twitter Statistics Reveal 100M Monthly Active Users & 250M Daily Tweets #w2s," *CMS Wire*, October 18, 2011, http://www.cmswire.com/cms/social-business/new-twitter-statistics-reveal-100m-monthly-active-users-250m-daily-tweets-w2s-013103.php.

[294] "Statistics," *YouTube*, http://www.youtube.com/t/press_statistics.

[295] K. J. Mason, "Social Media Statistics and Facts of 2013 [INFOGRAPHIC]," *Growing Social Media*, June 10, 2013, http://growingsocialmedia.com/social-media-statistics-and-facts-of-2013-infographic/.

[296] Jeff Bullas, "20 Stunning Social Media Statistics Plus Infographic" (blog), September 2, 2011, http://www.jeffbullas.com/2011/09/02/20-stunning-social-media-statistics/.

[297] "Media Reaction to London 2012 Olympic Opening Ceremony," *British Broadcasting Corporation*, July 28, 2012, http://www.bbc.co.uk/news/uk-19025686.

[298] Sufian Ullah, "Top 10 Most Read Newspapers in the World," *Click Top 10*, http://www.clicktop10.com/2013/07/top-10-most-read-newspapers-in-the-world/.

[299] Rolfe Winkler, "Facebook's Timeline for Success," *Wall Street Journal*, November 11, 2011, C10.

[300] Steve Lukasik, interview by Mike Daniels and J. Robert Beyster, September 18, 2008.

[301] J. C. R. Licklider, "Internet Pioneers," http://www.ibiblio.org/pioneers/licklider.html.

INDEX

Q
Q32 computer, 191
Quantum Computer Services, 104

R
Rand Corporation, 189
Raven Systems and Research, 35, 37
Regional IRs, 117–119
register.com, 148
RIPE (Rèseaux IP Europèens), 118–119
Roberts, Larry, 192–193
Rockwood, Steve, 165–166
Root name servers
  evolution of, 25–26
  function of, 23
  operators of, 125–126
  original design of, 25
  types of, 24
Roper, Bill
  on Internet awareness, 112
  leadership role of, 107–108
  on security, 178
Ruina, Jack, 192

S
SAG (Strategic Advisory Group), 65
SAIC (Science Applications International Corporation), See also, Network Solutions, SAIC acquisition of vi–vii, 7, 36
  administrative services by, 79–80
  culture of, 165–166
  DNS fees and, 74
  employee ownership of, 159–160
  growth opportunities in, 12
  Internet impact by, 107
  Internet policy and, 131

Internet purists and, 130
planetary organizational model, 187–188
A server responsibility of, 82
Telage at, 51
VeriSign payment, 164
The SAIC Solution: How We Built an $8 Billion Employee-Owned Technology Company, 12
Sbarbaro, Phil, 121, 147–148
Schrader, Bill, 30
Security issues, 89–90, 177–178
Semi Automatic Ground Environment (SAGE) air defense system, 191–192
Shared Registration System
  idea for, 147–148
  protocol for, 147
  test organizations for, 148
  timetable for, 146–147
Skype, 181
Small Business Act, 38
Small Business Administration, 56
Sneeringer, Gerry, 129
Social media
  categories of, 183–184
  growth of, 182
  value of, 185
Spotify, 184
Sprint, 31
Sputnik, 190–191
SRI International, 16, 44, 80
Strawn, George, 27, 106
  on ARIN formation, 120–121
  on commercialization, 104–105
  cooperative agreement and, 50–51, 74
  on DNS fees, 73–75

4758247R00130

Made in the USA
San Bernardino, CA
05 October 2013